John Struthers

Memoir on the Anatomy of the Humpback Whale, Megaptera

Longimana

John Struthers

Memoir on the Anatomy of the Humpback Whale, Megaptera Longimana

ISBN/EAN: 9783337322922

Printed in Europe, USA, Canada, Australia, Japan

Cover: Foto ©berggeist007 / pixelio.de

More available books at **www.hansebooks.com**

ON SOME POINTS IN THE ANATOMY OF A MEGAPTERA LONGIMANA. By JOHN STRUTHERS, M.D., *Professor of Anatomy in the University of Aberdeen.* (PLATES I. and II.)

PART I.

HISTORY AND EXTERNAL CHARACTERS.

Parts Noticed and Order.

1. HISTORY.—This Megaptera, a male, 40 feet in length, had shown itself in the Firth of Tay, off Dundee, for five or six weeks before the end of December 1883, when it was at last fatally wounded. After about a fortnight it disappeared for eight or ten days, then reappeared in the Tay, and during the three weeks before its death disported itself freely in sight of the inhabitants of Dundee, up and down the river, going up as far as the docks. The attraction to the Tay was believed to be the young herring with which the firth abounded at the time.

From the newspaper accounts of the appearance and movements of this whale in the firth, I had inferred that, although very rare on British coasts, it must be a *Megaptera longimana.* As described to me by witnesses who watched its movements, it rose out of the water seemingly for two-thirds of its length, almost perpendicularly, flapped its enormous paddles, and then fell to one side, causing great commotion in the smooth water.

This movement it was seen on one occasion to repeat three times. The movement was described as resembling the leap of a salmon, but slower. The intervals between the blowings were generally about two minutes, never more than five minutes. A stream like a spray fountain went up for, it seemed, 15 to 20 feet, at first straight up and then broke. The blow-hole part was not visible above water. When it rose ordinarily the back was seen first, then the dorsal fin; in disappearing, the dorsal fin was the last seen; neither the tail nor the paddles were shown.

When at last successfully harpooned it showed great strength and endurance for twenty-one hours, when the line parted, but it had been mortally wounded.[1] This was on New Year's morning, 1884. A week afterwards the carcase was observed by fishermen off Bervie, on the coast of Kincardineshire, floating so high as to be visible 6 miles off. It was towed into Stonehaven harbour on January 8, and beached there.

My first observations and measurements were made as it lay on its back at Stonehaven, and photographs were taken, from one of which fig. 1 is taken. On the day after it was beached, the carcase, the property of the fishermen who found it, was exposed by public sale and purchased for a large price by Mr John Woods, oil merchant, Dundee, with a view to exhibition.

---

[1] Some particulars of the endurance may be interesting. After the first harpoon, which was thrown and went in at the shoulder, it swam quietly, rising at intervals of two minutes to blow, but the vapour was reddish. After a second harpoon, which was fired, took effect, it made vigorous efforts, threw the tail in the air, lashed the water furiously and darted about in different directions. Volumes of blood were now thrown up, colouring the surrounding water. It had at first to drag two six-oared rowing boats and a steam launch, and, four or five hours afterwards, a steam tug was added. With this heavy drag it swam wildly about, on one occasion rising under one of the boats and lifting one end of it out of the water. Hand-lances were driven 3 feet deep into it, and blood spouted from the wounds. Two of the harpoon lines parted, but the steam tug and the two rowing boats were dragged out to sea by the remaining line, north to near Montrose, south to near the mouth of the Firth of Forth, then north again. At daylight a 4-feet-long iron was fired into it, also a couple of marling-spikes, and a number of iron bolts and nuts. About twenty-one hours after being harpooned it showed signs of exhaustion, turning from side to side and lying level on the water, but shortly revived and again held on; in half an hour the line parted, some way south of the Bell Rock, and the whale was free. The cruelty, which one cannot but recognise, of this long chase was largely owing to deficiency in modern appliances of attack.

The carcase was removed the same night to Dundee, tugged by a rope attached to the tail.[1]

2. DISSECTION OF THE CARCASE.—After it had lain a fortnight for exhibition, I was allowed to make a dissection of the carcase, in which I was assisted by Mr Robert Gibb and Mr George Sim, of Aberdeen, and by several Dundee whale-fishers.[2] The carcase having been turned on the back, my first care was to remove a large portion of the abdominal wall, in its whole thickness, from the umbilicus to behind the anus, and of sufficient breadth to include the pelvic bones and rudimentary hind limbs. This half-putrid mass was pickled and sent on to Aberdeen to be dissected at leisure. On looking for the viscera they were found to be so decomposed as to be mostly unrecognisable, reduced along with the muscles to a pulp into which the whale-fishers went knee-deep. We tried to preserve the heart, but our hands went through it. Our attention was therefore directed to securing the bones, some of which came out already detached from the soft parts. The vertebræ, except those of the narrow part next the tail-fin, the sternum, ribs, and hyoid, were removed, and sent on to my macerating troughs at Aberdeen.[3]

On August 7, fully seven months after the death of the whale, I went to Dundee to complete the removal of the bones,

[1] When lifted out of the water in Dundee harbour with the steam crane, by a chain round the tail, high in the air, the tongue and some other soft parts, and the cervical vertebræ, fell out by the mouth into the water. The vertebræ were recovered. It was then placed on the belly on railway lorries, and dragged by eighteen powerful horses along the docks, and, after various mishaps, reached Mr Woods' yard, where it was placed for exhibition.

[2] I may here remark, in apology for delays and shortcomings in my observations of the external characters and internal structure of this Megaptera, that everything had to be subordinated to its exhibition at Dundee and then at other towns. The dissection was not only late (January 25 and 26), but was attended by difficulties and by unusual accompaniments. When we arrived to perform it, we found that the astute proprietor had announced a special admission, adding the attraction of a band of music, and I may add we had a snow-storm which drove us off from time to time.

[3] The remains were then prepared for exhibition by the proprietor, Mr Woods. The putrid soft parts having been scooped out, and the remaining soft parts prepared with antiseptics, a wooden backbone was introduced, wooden bars supplied the place of ribs, and the body was stuffed and stitched below into proper form. The embalmed whale, thus wonderfully restored in form and much lightened, was exhibited during the next few months in various towns, first in Aberdeen, then in Glasgow, Liverpool, and Manchester, again in Glasgow, in Edinburgh, and finally again in Dundee.

in which I was assisted by Mr Robert Gibb, Aberdeen, and Dr
Greig, junior, Dundee. The blubber in being cut in pieces was
seen to average about 3 inches in thickness and was in a fair
state of preservation. The skull and remaining bones were
sent on to Aberdeen, the paddles entire, so that I could dissect
them carefully. The skeleton was presented by Mr Woods
to the Dundee Museum, where it will be finally placed.

3. MEASUREMENTS—   ft. in.

Length, from point of lower jaw to cleft of
    tail, straight, ... 40 0
Pectoral fin, length along lower border, . . 12 0
       „ greatest breadth, . . 2 8½
Dorsal fin, height of fin proper, . . . 0 7
       „ height of entire elevation, . . 0 11
       „ distance from cleft of tail, . . 12 4
Tail fin, width between the tips, . . . 10 6
       „ extreme width, straight, . . . 11 4
       „ greatest antero-posterior breadth, . 3 0
Mammillary pouch, in front of anus, . . 2 0
Projection of lower jaw beyond upper, . . 1 1
Point of lower jaw to angle of mouth, straight, 9 4
Same to below angle of mouth, along the
    curve, ... 10 4
Point of lower jaw to pectoral fin, . . 14 1
Eye to eye, centre of, over the top, . 7 10
Distance of latter line behind blow-holes, . 1 2
Ear-hole behind posterior canthus of eyelids, . 1 5
Whalebone, largest plates, length, . . . 1 8
       „ largest plates, greatest breadth, . 0 5

4. SIZE.—The 40 feet straight measurement, to the mesial
cleft of the tail, was increased only 1 foot by following the
curves on the side, only half a foot when taken over the belly
To the most posterior part of the tail-fin gave 9 inches more
than to the median cleft.[1]

[1] The common statement that Megaptera when mature may reach a length
of nearly 60 feet, appears to rest on the authority of Captain Holböll.
Fabricius had put it at 50 to 54 feet, but even that would appear to be an unusual
size. Mr A. H. Cocks (*The Zoologist*, 1884, 1885, 1886, and 1887) gives much
interesting information in regard to finners, obtained during his visits to the

5. PECTORAL FIN.—The great length of the pectoral fin, or paddle, is the most striking character of Megaptera among all the whalebone whales. The following table shows the length of the paddle in proportion to the length of the entire carcase in those I have had the opportunity of measuring :—

	Length of Whale.		Length of Pectoral Fin.	
	ft.	in.	ft.	in.
Balænoptera musculus, Wick, 1869, . . .	65	6	8	8
Balænoptera musculus, Peterhead, 1871, . . .	64	0	7	8
Balænoptera musculus, Stornoway, 1871, . . .	60	6	7	1½ [1]
Balænoptera musculus, Nairn, 1884, . . .	50	0	5	11
Balænoptera borealis, Orkney, 1884, . . .	35	0	4	7
Balænoptera rostrata, Aberdeen, 1870, . . .	14	6	2	2
Balænoptera rostrata, Bervie, 1877, . . .	16	0	2	3
Balæna mysticetus, Davis Straits, 1873, . . .	48	0	8	6
Balæna mysticetus, Davis Straits, 1874, . . .	35	0	5	6
Megaptera longimana, Dundee, 1884, . . .	40	0	12	0

These measurements are from the head of the humerus to the tip of the paddle. That is little less than the measurement along the inferior (radial) border when the fin is still attached to the body. The measurement along the ulnar border is con-

Fin-Whale Fisheries, where large numbers of the various finners are killed. The common statements in regard to the lengths attained among the other finners are borne out, but not so in regard to Megaptera. Plenty of instances are mentioned of the Blue Whale (*B. Sibbaldii*) reaching from 70 to 80 feet, or more, giving averages of 75 and 79 English feet ; of the Razorback (*B. musculus*) from 60 to 70 feet, with averages of over 64 feet ; and of the Black Whale (Rudolphi's rorqual, *B. borealis*) from 40 to 45 feet, some nearly 50 feet. Of the Humpback (*Megaptera longimana*) the largest averages of length were Captain Horn's, being, on a take of 6 males, 41½ English feet (the longest 53 feet), and of 2 females, 46½ feet (the longest 48 feet). Mr Cocks remarks—"The average for all the males whose length is given above is under 35½ English feet, while that of the females is just over 40½ English feet. The number of Humpbacks taken that year (1886) is given at 94 ; of the Blue Whale, 152 ; of the Razorback, 646; of Rudolphi's rorqual, 62 ; total 954, by 39 whalers." Mr Cocks says of the fishery of 1885— "Captain Berg told me that he had this season captured the biggest Humpback he had hitherto seen. It was a female, and measured 50 Norwegian feet (52 feet English) in a straight line (measured as Dr Guldberg had directed)." While the state of some parts of the skeleton will sufficiently show that my Megaptera was not full-grown, it would appear, from the lengths given by Mr Cocks, that a 40-feet male Megaptera is not so far from being adult as the nearly 60 feet statement might have led us to infer.

[1] In this *B. musculus* the length of the paddle would have been a little more but for a mal-development near the point.

siderably less. In the 50-feet-long B. *musculus* the length was 4 feet 2 inches along the ulnar border, 6 feet along the radial border, and after removal the measurement from the head of the humerus was the 5 feet 11 inches given in the table. In this Megaptera the measurements were:—inferior border, 12 feet; superior border, 10 feet 3 inches; along the middle, straight, 11 feet 8 inches; from the head of the humerus, after removal of the fin, 12 feet. The measurement at the shorter border would give the pectoral fin of Megaptera a proportion of more than a fourth of the total length of the carcase; that along the inferior border as rather nearer a third than a fourth. In the other great finners the proportion may be put at about an eighth, taken on the lower border.

The paddle in Megaptera is also of greater breadth than in other finners, though not than in Mysticetus.[1] In the 64-feet-long B. *musculus*, the greatest breadth was 19 inches, in the 50-feet-long one, 16 inches. In this Megaptera it is $32\frac{1}{2}$ inches. The breadth is, however, not greater in proportion to the length in Megaptera than in the full-grown B. *musculus*, but rather less.

Another external character of the paddle of Megaptera is the very undulating anterior border, showing two greater and seven lesser nodes (see fig. 1), the causes of which are made clear by the dissection. A few lesser undulations are seen on the ulnar border towards the tip.

6. DORSAL FIN.—The form is shown in fig. 2. There is what may be called the fin proper and the basement, a low elevation from which the fin proper rises abruptly behind and gradually in front. The basement is marked off only by a gradual elevation from the general contour of the back. It extends from about 2 feet behind the tip of the fin proper to about 4 feet in front of it. Height of fin proper 7 inches, of the entire elevation 11 inches. Thickness at the middle of the fin proper, $1\frac{3}{4}$ inches; at base of fin proper, 3 inches; of basement at its lower part, 12 inches; at 2 feet in front of fin proper, and at mid-height of basement there, 4 inches. In Rudolphi's

---

[1] Mr Robert Gray informs me that his father, Captain David Gray, this summer caught the largest Right Whale (*Balœna mysticetus*) he has ever taken, a female 57 feet long; length of the pectoral fin, measured along the middle of the outstretched fin, on the inner surface, 8 feet 2½ inches, greatest breadth 5 feet 1 inch.

diagram [1] the dorsal fin is not very like this one, stands up more, and the point and both margins are different. In the small figure given by Eschricht (p. 152, fig. 48), the notch is much less marked than in this one. As seen in my figure, the anterior slope is a little concave on the basement, and becomes convex on the fin proper. No exact spot could be fixed on here for the commencement of the fin proper as distinguished from the basement, but the level of the notch behind determines that to the eye. The point is some way behind the top, with a fall of about 1 inch. The posterior border below the point is convex down to the bottom of the notch. This, with the gentle concavity of the posterior slope of the basement, renders the notch pretty sharp, more so than in the higher and more recurved fin of *B. musculus*.[2]

As to situation, the distance from the cleft of the tail to the notch of the dorsal fin was 12 feet 4 inches of the 40 feet. In my 64-feet-long *B. musculus* the distance was 15 feet 8 inches (height of fin 15 inches, length at base 24 to 26 inches). This would place the dorsal fin further forwards in Megaptera than in *B. musculus*.

7. TAIL-FIN.—The form of the tail-fin is shown in fig. 3. Its greatest antero-posterior breadth was 3 feet, only 3½ inches more than the breadth of the pectoral fin; its total width less than the length of the pectoral fin by 8 inches. The statement of the depth of the median cleft, commonly said to be deep, will depend on where the measurement is taken. From between the neighbouring convexities, about 7 inches out, the depth is 3½ inches; from between the first prominent serrations, about

---

[1] *Abhand. könig. Acad. der Wissenschaften*, Berlin, 1829, Taf. v. fig. 1.

[2] I have noted these points particularly on account of the question of the origin of the name "Humpback" for this species. "Les mégaptères ont une bosse sur le dos à la place d'une nageoire"—"une véritable bosse dépendante de la peau," says the eminent cetologist P. J. van Beneden. There was nothing in the appearance of the back of this Megaptera to suggest to us the appropriateness of the name Humpback. That, however, will depend partly on the idea one associates with the word humpbacked. The name may have arisen rather from the rounded back Megaptera shows above water, as long ago suggested and figured by Eschricht (*Untersuchungen über die Nordischen Wallthiere*, p. 152, fig. 48) :—"Der Name *Humpback* scheint übrigens nicht nur von der Rücken-flosse, sondern eben so wohl von dieser Krümmung des Rückens beim Unter-tauchen." The term, though rather misleading as to the true form, is a convenient one to the whale-fishers.

16 inches out, the depth is 5 inches; from the most projecting part of the posterior border, more than half-way out, it is 9 inches; and from between the recurved tips the depth is 17 inches. The cleft proper is indicated in the first of these measurements, but some may have taken it at the third.

From tip to tip, straight, the tail-fin is 10 feet 6 inches; between the extreme edges in front of the recurved tips, 11 feet 4 inches.[1] The anterior border, after the neck, is convex throughout and very much bent back and also inwards, towards the tip. The falling in is 5 inches, so that the broadest part of the fin is in front of the tip. This great bending back extends on about 18 inches of the anterior border and 9 inches of the posterior border. The latter undulates; the broadest and most projecting convexity is external to the middle of each half; the concavity between that and the convex boundary of the median cleft is shallow; the concavity next the recurved tip is a deep bay, about 9 inches deep externally and 2 feet in width. The inward direction towards the blunt tip is mainly on the anterior border, but even the posterior border is here directed a little inwards. After the smooth convexity bounding the median cleft, the whole posterior border, out to the tip, is serrated; about twenty serrations may be counted (on each half of the fin), some large, some small, some sharp, some rounded. The antero-posterior breadths are, in inches, near the median ridge, 36; at junction of inner and second fourths, the same; midway out, $32\frac{1}{2}$; at junction of outer two-fourths, 24; at nine inches from the tip 7 inches transversely.

8. SURFACE OF THE ABDOMEN.—The *umbilical fissure*, or groove, 14 inches in length, begins immediately behind the plaitings of the skin. For 8 inches it is a well-marked elliptical fossa, deepest at the anterior end and feels hard at the middle.

The *preputial opening* is 3 feet behind the fore-end of the umbilical fissure. The epidermis is whitish here and for 4 or 5 inches back. From the preputial opening to the anus there is a groove in which the skin is soft. At the prepuce, at the white part, the groove is at first 4 inches broad and deep, over

---

[1] Straight from extreme to extreme is the usual way of measuring the tail-fin, but a truer method would be to measure straight from the extreme to the middle of the median ridge, giving the width of each half. Here that is 6 feet 3 inches, giving $12\frac{1}{4}$ feet as the axis of the entire tail-fin.

the cavity for the penis; it then narrows backwards to the mammillary pouch, and is narrow from that to the anus.

Mammillary Pouch.—This interesting part in this *male* Megaptera is situated 1½ feet behind the preputial opening and 2 feet in front of the anus.[1] The following is the arrangement (see fig. 4, natural size):—The opening of the marsupium, elliptical in form, is from 1½ to 2 inches in length, more sharply marked behind, grooved for ½ inch at the fore-end; breadth ¾ inch; the margins soft from the looseness of the subcutaneous tissue. The black colour continues

[1] Referring to Pallas having first noticed the presence of mammillæ in the male cetacean, in the Beluga, Eschricht mentions particularly that their presence in male whales, fœtal and adult, has been well known to him. There is no reason why the milk glands should not be present in male whales as well as in male land mammals; they cannot be more functionless in the former than they are in the latter, or than they are in man. The point of interest is how these significant rudiments are variously disposed. Eschricht found them present in all male cetacean fœtuses—"an der Mittellinie des Bauches ein Paar kleine schlitzenförmige Öffnungen," and that in the male porpoise (*phocœna*) " sie nach aussen hie zu einer einfachen Öffnung verschmolzen sind" (*loc. cit.*, 1849, p. 83). Professor Flower (*Proc. Zool. Soc.*, 1865, p. 701) found the arrangement in an adult male *B. musculus* to be that of two fissures, about 10 inches long, 1½ inch deep, and 3 inches apart, slightly converging posteriorly, each containing a nipple. The two-fissure arrangement, one on each side, more resembles that of the female. John Hunter described the position and structure of the mammary glands and nipples in the female cetacean (*Phil. Trans.*, 1787), and figured the nipples in their fissure in a 17-feet-long *B. rostrata* (Table xxi.). He describes the nipple as lodged in a sulcus on each side of the opening of the vagina, surrounded by loose texture, and, external to this, another small fissure, "which I imagine is likewise intended to give greater facility to the movements of all these parts." As these parts in my 14½-feet-long *B. rostrata* (1870) are preserved, I may here mention that they differ from Hunter's figure in the accessory fissure, instead of rather shorter, being much longer than the mammillary fissure. Length of accessory fissure 5½ inches on the left side, on right side 4½; length of the mammillary fissure, 2 inches on both sides. The right accessory fissure passes as far back as the mammillary fissure, the left ¾ inch farther back. Breadth of skin between the two fissures 1 inch, being about the same as that between the vulva and the mammillary fissure. The nipple lies behind the middle of the fissure, concealed in it, ¼ inch or more from the surface, is flattened and now about ¼ inch in height, and is surrounded by a deeper and softer part of the fissure. An aperture in the summit admits a crowquill, and a little way along the duct, in the nipple, two or more apertures are seen. The accessory fissure is deeper than the mammillary fissure. In my 16-feet-long *B. rostrata* (1877) the accessory fissure has not been preserved. The mammillary fissures are each 2 inches in length. The middle ½ of the fissure forms a special fossa round the base of the nipple, 1 inch deep from the surface, thrice as deep as the anterior and posterior parts of the fissure. The flattened nipple is ¼ inch in height.

to a little within the edge of the opening, where the walls of the pouch and all the parts within it become white or cream-coloured. Plugging the mouth of the pouch is a large soft projection like the pulp of the thumb, as if a distended septum, but the lining membrane of the pouch dips in half an inch before and behind it, the depth of the pouch being about $1\frac{1}{4}$ inch at other parts. The nipple is brought into view by pushing aside the septal plug or the outer wall of the pouch, as seen in the figure. The nipple, flattened sideways, projects like a thick tongue, $\frac{3}{4}$ inch in height, $\frac{1}{2}$ inch in breadth. On the outer side of the nipple, a little way from the summit, is a large aperture, admitting a goose-quill, shortly within which two apertures are seen, as if the main duct there divided. Into one of these a middle-sized probe passed readily for $1\frac{1}{4}$ inches. The whitish epithelium on the median plug and nipples was about $\frac{1}{26}$ inch thick, and when this was removed, the cutis vera on the summit of the plug and nipples presented tufts instead of the fine filiform processes which their other parts showed. In one of the photographs, taken the day after the whale was beached, the median plug can be recognised bulging moderately in the mouth of the pouch.

9. THE PLAITINGS OF THE SKIN.—These are much broader and consequently fewer than in other finners. The breadth is about $4\frac{1}{2}$, or maybe 5, inches. The furrows, after a few inches, have gained a depth of 1 inch and reach a depth of 2 inches, some $2\frac{1}{2}$, and are dark to the bottom. The number of plaits is about twenty-four. They extend from below the lower jaw to the front of the belly, ending there on a line drawn from 2 feet behind the axilla to the umbilicus. Two of the furrows, the 2nd and 7th below the axilla, are not continued forwards; the same of the 11th, but it is longer. The median furrow is not continued so far back as those next it. There is a short (13 inches) azygos furrow to the right side of the umbilicus, which if continued forwards would have split the median plait. The line seen in fig. 5 above the shoulder is not one of these furrows, but only a fold of the skin. The system of furrows begins below the side of the mandible, below the labial groove, by two furrows, closed at each end, as shown in fig. 5. The furrows of the throat run forwards close to the mandible, within 2 inches of it,

towards the symphysis, within about 5 inches at the sides. When the carcase lies on the belly, plaitings are thus visible below the mandible. Where the skin turns in from this to below the throat the plaitings present white patches, and this part is seen to form a projection, like a second chin, in figure 1 when the carcase lies on the back. A little behind this two of the furrows terminate, two of the plaitings having bifurcated backwards at the fore part of the throat. All of the long furrows are not continued throughout the length of the plaited area. Thus, the second and third furrows below the axilla are confluent backwards at about 2 feet in front of the axilla; a furrow at about half-way between the axilla and the mesial line ends opposite the axilla; and the fourth furrow below that one, mesial or nearly so, stops about 2 feet farther back, and is the lower limb of a furrow which has bifurcated backwards about 5 feet in front. The furrows seen in figs. 1 and 5 are exactly as in the photographs. I could not ascertain whether the furrowing is quite symmetrical.

10. DERMAL TUBERCLES ON THE HEAD (see fig. 5).—These large dermal tubercles rise to a height of 1 inch, one or two of the posterior of the median row to $1\frac{1}{2}$ inch. They are elongated antero-posteriorly. All are soft when pierced. On the upper jaw the median row has seven tubercles, at distances varying from 6 to 12 inches, which are connected by a low median ridge. The lateral row has eight on the right side, eleven on the left, but arranged in pairs except the foremost and hindmost, and are therefore at longer intervals than in the median row. On the mandible there are, along the side, six, the two hindmost low down, the other four arranged in a row high up; and close to the symphysis there are six on each side, forming an irregular cluster, placed mostly below the middle of the symphysis, the tubercles projecting like the end of a hen's egg, some twice that size. There are thus twenty-six great tubercles on the upper jaw, twenty-four on the lower.

11. HAIRS.—Most of the hairs seen on the lower lip had disappeared before I could attend to them. The eight which I took out vary from $\frac{3}{4}$ to 1 inch in length, are white and pretty stiff. They were readily seen by standing sideways to the tubercles. They projected $\frac{1}{4}$ to $\frac{1}{2}$ inch and came out easily

between the finger and thumb. It had been noticed before that some of them projected ¾ to 1 inch, but whether this was natural or owing to the hair coming out I cannot say. If the piece of black epidermic sheath adhering is to be taken as marking where the follicle began, one of the eight I have preserved must have projected for ½ inch. Our inability to find hairs on the upper jaw was not surprising, as the brushing and usage this part had received had already removed the epidermis. Those I took out grew from the tubercles at the symphysis, but one was found two feet back from the symphysis, on the second lateral tubercle, projecting about ¼ inch. It is an interesting question in what relation the tubercles and the hairs stand to each other.

12. ADAPTATIONS OF THE JAWS.—The projection of the lip of the mandible beyond the upper jaw is, at the front 13 inches; at the side, before the labial groove begins, 13½. The thickness of the soft tissue (lip or gum) forming the upper edge of the mandible was, at 1 foot from the symphysis, 2 inches; at the side, where the labial groove begins, about 6 inches.

Labial Groove (see fig. 5).—At 3½ feet from the symphysis, being about one-third of the distance along the side of the mouth, the lip bifurcates to form a deep broad groove, the inner boundary of which is the continuation back of the jaw, the outer border cutaneous. This groove deepens and broadens backwards to a breadth of 15 inches as a deep groove, and, becoming gradually shallower and broader (reaching a breadth of 30 inches), it is lost on the surface on a line drawn from the angle of the mouth downwards and forwards. The furrow seen below the shoulder in fig. 5, as if continued from it, is not a continuation of it. The whalebone range descends within the mandible, and the water escaping from between the plates will be conducted backwards along this labial groove. I am not able to say whether this great spill-water groove is in any way peculiar to Megaptera.

13. CUT-WATER.—At the point of the mandible, which is blunt (transversely 8 inches, vertically 7 inches), a median ridge goes down to a median projection, placed like a prow or cut-water. The height of this cut-water is 14 inches; breadth, 2 inches; amount of projection, 4 inches, the lower third

sloping backwards. The vertical measurement of the symphysis and cut-water together is 21 inches.

14. THE WHALEBONE.—The largest plates are 20 inches in length. At the front for 3½ inches there are no plates, only about ⅛-inch-thick bundles at the gum, breaking up into hairs, but the two sides quite meet at the mesial line. The first plate has, on the outside, 2 inches of plate proper and 2 of fringe. Entire length of the range, 8 feet 2 inches at the top, at the fringe 9 feet. The following is the length of the plates, in inches, at different parts:—At 2 feet from the front, 11½; at 4 feet, 7; at 6 feet, 20; at 7 feet, the same; at 8 feet, 10 inches. The range ends behind in short bundles with fringe. The length of the hairy fringe below, along the range, is, at 6 inches from the front, about 2 inches; at midway back, about 3; at the longest plates, 5; behind this the fringe shortens to 4 inches. The backward obliquity of the fringe is greater than that of the plates. The most anterior plates are nearly vertical; when the middle is reached the slope backwards is equal to the breadth of nine of the plates as they appear externally. At 20 inches from the back the plates are nearly vertical; then they become vertical, and at the very back seem to slope a little forwards. The greatest breadth of the longest plates is 5 inches. Breadth of roof of mouth between the whalebone ranges is, at 1 foot from the front of the whalebone, 6½ inches; at 2 feet back, 8 inches; at 3½ feet back, 9 inches.

In colour, the whalebone on the outside was black, except along the front 12 inches where it was partly white, mottled, but differing in this respect on the right and left sides. On the left jaw here, at 6 inches from the mesial line, fifteen plates are quite white on their anterior half but black on the palatal half. Some near these, again, have the anterior edge black and the rest of their surfaces white. Viewed from the palatal aspect, the whole matting of hairs was whitish. The words in my note-book are "white, dirty-white, or yellow-white." Now, in 1887, after three years' exposure, though washed clean, that description could not apply. The colour of the hairy matting now is dirty-brown mixed with brown-black. The hairs are fully 4 inches in length, some 6 inches. The hairs of the fringe are thick and stiff, like bristles, compared with those of my

50-feet-long *B. musculus*, but the much finer hairs of the matting on the palatal aspect do not differ in thickness in these two whales.

15. BLOW-HOLES.—The length of the blow-holes is 11 inches; distance between hinder ends, 9 inches; between fore ends, 3 inches. They are a little convex towards each other. The median fissure has a depth of 1 inch at the middle. There is an elevation of the head here, rising about 3 inches, on the hinder slope of which the blow-holes are situated.

16. EYE, AND EAR-HOLE.—The *eye* is placed very close behind and above the angle of the mouth. The *ear-hole* is 17 inches behind the posterior canthus of the eyelids, and 2 to 3 inches below the level of the eye (see fig. 5). The epidermis being off, I could not ascertain whether there was any change of colour here.[1] The tissue immediately around the aperture is softer, so that a shallow depression can be made by the end of the finger. This will facilitate collapse of the meatus. The aperture admits a rather small-sized uncut goose-quill. In form it is ovoid antero-posteriorly, the anterior end sharp-edged, the posterior and narrower end grooved, the groove prolonged for about the same length as the foramen. This form of the aperture of this mammalian vestige may be an adaptation to forward swimming. Water in the meatus will be less disturbed. The quill goes straight into the meatus, at right angles to the axis of the body and head, firmly grasped, for 1 inch in the right, for 2 inches in the left.

17. COLOUR.—As it lay on the back, the day after it was beached, exposed by the retiring tide, the whole carcase appeared black, except the under surface of the tail-fin and of the breast-fin, whose snow-white appearance formed a striking contrast. The photographs taken on the same day show some patches of white on the throat and chest, notably on the plaitings below the chin, as they turn in below the mandible, and a few less abrupt streaks and smaller spots here and there along the chest. I had the opportunity of examining the white marks on the abdominal wall more carefully. As if bounding the perinæum, there was on each side, 5 or 6 inches out from the

---

[1] In my quite fresh 14½-feet-long *B. rostrata*, there was a white line leading backwards from the ear-hole for 9 inches.

mesial line, a white streak, like a chalk line on a blackboard.
It began 2 to 3 inches in front of the mammillary pouch, and
was seen as far as the epidermis was present, which was for 12
inches. The white went through and through the epidermis,
and a corresponding groove was present in the cutis vera,
traceable as far back as 10 inches behind the anus on the left
side, on the right side only for 12 inches behind the mammillary
pouch. The papillæ of the cutis were shorter and finer on the
groove. Besides these perineal lines and the white at the
prepuce, there were, near the mammillary pouch and forward to
the umbilical region, white spots like hailstones, and towards
the umbilical fissure a few white streaks. These streaks were,
on the left side, some 6 inches from the mesial line, and linear,
and on the right side a row of spots corresponded to one of the
streaks of the left side. These white streaks went through and
through the epidermis, and corresponding grooves were present
in the cutis vera.[1]

[1] *Variations in the Colour of Megaptera.*—The colour of the *outer surface of the
pectoral fin* in this Megaptera was stated to me, by observers who had good views
of the whale as it sported in the Tay, to be black. But where it lay on its belly
at Dundee that surface was variously stated to me to be white, to be black, and
to have black patches. When I went to dissect it at Dundee the epidermis was
off. Statements of the colour of parts of a whale when not fresh, and not made
by an experienced observer, are not reliable. When the epidermis is off, the
cutis vera is at first white or cream-coloured, like the skin of a well-washed
white pig; then under exposure for some time to the air it becomes bluish, and
on being scraped, the cream colour is restored. The acquired bluish colour
appears to be on the fine hair-like papillæ of the cutis vera. In regard to the
question of the colour of the *outer* surface of the paddle of *Megaptera longimana*,
Eschricht says (p. 147)—"Die Brustflossen aber sind an beiden Flächen rein
weiss." From the account given by Mr A. H. Cocks (*loc. cit.*) of a number of
Humpbacks he examined on shore, it would appear that there is considerable
variation in the colour of this part. He notes, 1884—(1) Male, 40 feet, paddles
black on the outer side, white on inner side, the black extending to round the
borders, "with an occasional blotch of black, and two or three black rings" also
on the inner side; length of paddle to head of humerus 11 feet 4 inches, greatest
breadth 3 feet 2 inches. "The throat with the furrows and nearly the whole of
the under side was white." (2) Male, 44 feet, outer surface of paddles black on
the proximal quarter only. Length of paddles 15 feet (measured to skin of
axilla about 13 feet 9 inches), greatest breadth 3 feet 7 inches. (3) Male, 30 feet,
paddles "only black on the upper side a little way down from proximal end."
In 1886—(4) Male, 35 feet, outer side of paddles "black for only a very short
distance at the proximal end." (5) 41–42 feet, outer side of paddle black only on
proximal quarter, "the black extending down the anterior edge, with a few
small irregular black marks lower down." (6) Male, 42 feet, "very little black
on the outside of the flippers, including a narrow rim along the hinder edge."

18. Skin and Blubber.—The epidermis on the part of the abdomen, which I had the opportunity of examining carefully, was from $\frac{1}{6}$ to $\frac{1}{8}$ inch thick. The cutis vera, when denuded of epidermis, had the cream colour already noted. The blubber at the fore part of the carcase was 4 inches thick, at the back part scarcely 3 inches, gradually diminishing from the middle back to the anal region where it was only $2\frac{1}{2}$ inches thick.

19. Parasites.—When beached at Stonehaven, parasites, seemingly of the usual kind (Diadema) found on Megaptera, were seen, but they were taken away by visitors. I noticed some on the distal part of the pectoral fin (inner surface) and some on the abdominal wall. The marks of the latter remain on the part of the wall which was preserved. One is seen on each side of the mammillary pouch ($1\frac{1}{2}$ to 2 inches from it)— large oval excavations, $2\frac{1}{2}$ inches by $1\frac{1}{2}$ inch; depth at the middle, one $\frac{1}{4}$ inch, the other nearly $\frac{1}{2}$ inch; the sloping edge of the epidermis is white at some parts; the cutis vera is smoothly excavated. A third is seen close to one of these; and about 1 foot forwards, on each side of the prepuce, 2 to 3 inches from the mesial line, three such excavations are present.

Length of flipper 12 feet 11 inches. This Megaptera was "entirely black on the belly, but nearly the whole of the thorax (i.e., chest and throat) was white, the chin being black, with a few white flecks." So experienced an observer as Mr Cocks was not likely to be misled by skin denuded of epidermis. The two last-mentioned whales had been brought into the factory only during the night before. It would seem, therefore, that, while the whole outer surface of the paddle may be black, the black is usually confined to the proximal fourth or less. This will be the part most visible above water, which may account for the impression of those whose observation was confined to the living animal that the outer surface of the paddle is black. The white colour of the under surface of the throat and chest in Nos. (1) and (6) of Mr Cocks' specimens, above noted, is a remarkable variation. My impression is that the mesial part on the under surface of the tail-fin, in my Megaptera, did not partake of the white colour shown by the rest of that surface.

Captain David Gray, of Peterhead, who has had a very large experience in the Greenland whale fishing, informs me, in regard to irregular white patches, that wounds and scars, such as are caused by ice or rock scratches or fighting, heal white. Also that the natural white increases in extent and degree with age in Mysticetus.

20. EXPLANATION OF PLATES I. AND II.

Fig. 1. View of the whale as it lay on the back at Stonehaven, drawn by Mr A. Gibb, from a photograph by Mr George W. Wilson, of Aberdeen, taken on the day after it was beached there. The plaitings of the skin on the throat, chest, and abdomen are exactly given. The throat is concave, the tongue having floated out of the mouth with the retiring tide. White patches are seen on the projecting skin where it turns in below the mandible, and a few less marked patches of white on the throat and chest. The inner surface of the pectoral fin, and under surface of the tail fin, are seen to be white. The pectoral fin lay abducted, and was much foreshortened in the photograph. I have drawn it directed more naturally backwards, and rotated outwards, giving a full view of its length and breadth, and showing accurately the nodes on its lower (radial) border, nine in number, the first and fourth nodes the most prominent. A rapid fall is seen on the contour from an angle some way behind the anus, presumably from the longer chevron bones backwards.

Fig. 2. The dorsal fin, reduced to $\frac{1}{36}$th. The two kinds of shading represent the distinction between the fin proper and its basement. The position of the highest part and of the point, and the form of the notch are exactly given.

Fig. 3. The tail fin ; reduced to $\frac{1}{36}$th. The inward curve of the tips is seen.

Fig. 4. The mammillary pouch and the nipples ; natural size. The sides of the pouch are hooked out so as to bring the nipples into view, on each side of the median septal plug. The aperture of the primary milk duct is seen some way down on the outer side of the nipple. The full size of the aperture is shown on the right nipple.

Fig. 5. From a photograph taken at Dundee by Mr F. G. Roger, of Broughty Ferry ; reduced to about $\frac{1}{40}$, being the size of the photograph. On the upper jaw are seen the dermal tubercles, the median row single, the lateral row mostly in pairs. On the mandible the more posterior of the lateral tubercles on it are seen. The commencement of the plaitings of the skin is seen below the mandible. *a,* The cut-water ; *b,* the position of the blow-holes, on the hinder slope of an eminence ; *p.f.,* commencement of the pectoral fin ; *c,* the ear-hole, *e.g.;* the labial, or spill-water, groove. The narrow groove below the shoulder is not a continuation of it. The tongue is not seen, having fallen out.

B

PART II.

THE LIMBS.

(A) THE PECTORAL FIN.

1. PROPORTIONS AND FORM.—The measurements given in Table I. below show that the greater length of the paddle in Megaptera is obtained mainly in the digital part. In B. musculus the digital part is shorter than the arm and fore-arm together, in the proportion of about 2 to 3½ (2 feet 1½ inch against 3 feet 7 inches). In Megaptera the proportions are reversed, the digital part being to the arm and fore-arm together as about 5 to 7 (4 feet 10 inches against 6 feet 9 inches). The measurements in the Table show that the greater actual length of the arm and fore-arm in Megaptera than in B. musculus is contributed in about equal proportion by the arm and fore-arm. These proportions are seen by comparing figures 6 and 13, of the same length, the former that of Megaptera, reduced to

$\frac{1}{24}$, the latter that of the 64-feet-long B. musculus reduced to $\frac{1}{16}$.

Form.—The **paddle** of B. musculus presents, on the *radial border*, only a gentle elevation opposite the distal end of **the** radius, and thence sweeps in an even convexity to the tip. On the ulnar side there is the considerable but gradual **elevation** at the pisiform **cartilage, and thence** to the tip **the** even concavity, **giving the clean-cut edges** and elegant form, smooth-edged, curved, tapering, and pointed, of the paddle in that finner (fig. 13), in striking contrast with that of Megaptera (fig. 6). It is seen from that figure that most of the projections on the radial border of the paddle of Megaptera are caused by the great size and lateral projection of the cartilages of **the digital joints.** These nodes are nine in number, **and the reason for that num-ber is seen.** Node No. 1, **that nearest the** body, most abrupt on the proximal side, very great, and the longest, is caused by the projection of the end of the radius, and by the sloping away from it of the radial border of the carpus and of the index metacarpal. Nos. 2 and 3 are caused by the nodes of the index digit, No. 2 more gradual and not so high as No. 3. No. 4, a great hump-like projection, is caused by the expanded terminal cartilage of the index digit and the ending of that digit. It rises $4\frac{1}{2}$ inches beyond the general outline of the border, and is most abrupt on the proximal slope. The hollow between it and No. 3 is very marked, like that of a deep saddle. The remaining five projections are caused successively by the now exposed nodes of digit III. No. 5 is low but better marked than No. 2; Nos. 6, 7, and 8 are prominent, No. 7 the most so. The well-marked hollows between are rather wider than the mounds. No. 9 is situated at about 9 to 10 inches from the tip, is gradual and less prominent.

On the *ulnar border* there is the usual great but gradual elevation over the pisiform cartilage, its highest part a little proximal to the radial elevation. The border then sweeps on, gently concave, without undulation till about 20 inches from the tip, where two true projections occur, caused by the 6th and 7th nodes of digit IV. The second of these rises $1\frac{1}{2}$ inch, and is opposite the 8th prominence on the radial border. Between it and the tip occur four or five little undulations, but they are not prominences with a solid foundation, merely slight wavings of the soft parts. Digit V. causes no projection; the prominence formed by the pisiform keeps the skin away from the first two nodes of the digit, and its third node does not project on the ulnar side. The border does not come near digit IV. till opposite its last two nodes, which project very sharply on the ulnar side, and thus cause the two ulnar prominences above described.

The prominences on the paddle of Megaptera are, therefore, unlike those on the head, not dermal, but owing to the adaptations of the bones and cartilages. Their presence in Megaptera but not in B. musculus is owing to the very much greater expansion of the digital cartilages in the former. On piercing with the exploring needle at the prominences and between them, along the radial border, cartilage or bone was reached at from $\frac{3}{4}$ to 1 inch from the surface.

2. TABLE I.—*Measurements of the Pectoral Fin and its parts, of the Megaptera longimana, and of the 50-feet-long Balænoptera musculus.*

	Megaptera 40 feet.		B. musculus 50 feet.	
	ft.	in.	ft.	in.
Length of pectoral fin, attached,	...		...	
,, inferior border,	12	0	6	0
,, superior border,	10	3	4	2
,, along the middle, straight,	11	8		...
,, from head of humerus, when detached,	12	0	5	11
Breadth, greatest,	2	8¼	1	4
,, at middle,[1]	2	0		...
Length of humerus and fore-arm,	4	10	3	7
,, carpus,	0	5	0	3¼
,, digits,	6	9	2	1½
Humerus, length,	1	11½	1	5
,, greatest diameter of articular head,	1	0¼	0	8
,, breadth of shaft, at narrowest part,	0	9½	0	6¾
,, thickness at ditto,	0	6¾	0	4½
Radius, length along middle,	3	0	2	2
,, ditto without epiphyses,	2	8½	2	0

[1] The breadths of the paddle of this Megaptera at various parts were, as follows, in inches:—At 18 inches from the body, 23 ; at 30 inches from the body, the broadest part, 32½ ; at the middle, 24 ; at three-fourths from the body, 18 ; at 12 inches from the tip, 12 inches.

TABLE I.—*continued.*

	Megaptera 40 feet.		B. musculus 50 feet.	
	ft.	in.	ft.	in.
Radius, breadth at proximal end below the epiphysis,	0	7½	0	6¼
,, breadth at narrowest part,	0	5¼	0	3⅜
,, thickness at ditto,	0	4⅜	0	2½
,, breadth at wrist, above the epiphysis,	0	11¾	0	6¼
,, thickness at ditto,	0	6	0	3¼
Ulna, length,	2	6	2	0½
,, ditto without epiphyses,	2	3½	1	11
,, breadth at proximal end, including bony olecranon,	0	6¾	0	6½
,, breadth at narrowest part,	0	4¼	0	2½
,, thickness at ditto,	0	3	0	2¼
,, breadth at wrist, above the epiphysis,	0	7½	0	4¼
,, thickness at ditto,	0	3	0	1⅞
Width of interosseous space, at middle,	0	0¾	0	2
Breadth of radius and ulna together at wrist,	1	7	0	11¼
Carpus, length at middle,	0	5	0	3½
,, breadth, with pisiform,	1	11½	0	11¾
,, ditto without pisiform,	1	6½	0	9½

3. THE SCAPULA.—The scapula of Megaptera differs greatly from that of B. musculus in form, in being higher in proportion to its antero-posterior length, and in its upper border being more arched upwards, and in its thickness, but the chief difference is the absence in Megaptera of an acromion process, and in the rudimentary condition of the coracoid. A low thick spine is seen in much the same position as in B. musculus, marking off a very shallow prescapular fossa, about 2 inches in breadth; but this low spine disappears at 7 inches from the glenoid cavity, and there is not even a ridge where the acromion would have been placed. The *coracoid* projects only for 1 inch, but extends into the area of the glenoid cavity, forming the anterior and inner part of the cavity for a space of 3½ inches in width and 2 inches long-ways. The line of synostosis is still visible on the left scapula (fig. 8), but is nearly obliterated on the right scapula. The projection of the coracoid would be greater in the mature state, as on its low blunt end there is a triangular area 1½ inch by 1¾ across, which has been covered by cartilage.

The *proportions* of the scapula, actual and in comparison with those of B. musculus, will be seen from the measurements given in Table II. Its greater height in Megaptera, compared with its antero-posterior length, is well brought out by measurements 1, 2, and 3 in the Table. The line between the anterior and posterior angles divides the height equally, while in B. musculus only about one-third of the height is above that line. The anterior border is shorter than the posterior, measured straight from either the end or the middle of the edge of the glenoid cavity. In B. musculus the anterior border is longer than the posterior taken from the ends of the glenoid cavity, owing apparently to the presence of the large coracoid and acromion processes. The

4. TABLE II.—*Measurements of the Scapula, given in inches.*

	Megaptera.	B. musc. 50-feet-long.	B. musc. 60½-feet-long.
1. Antero-posterior length, right,	42	39	...
,, ,, left,	41	38½	48
2. Height, right,	30	22½	...
,, left,	29	22	27
3. To upper border from line of antero-posterior diameter,	15	9	11
4. To glenoid margin from ditto,	15	13	16
5. Anterior angle to anterior end of glenoid cavity, straight,	19	21	26
6. Posterior angle to post. end of ditto,	22	19½	24
7. Middle of glenoid margin to anterior angle,	24½	23½	29
8. Ditto, to posterior angle,	26½	23½	29
9. Glenoid cavity, length,	12¼	9	11
10. ,, breadth,	9¼	6¼	7½
11. Coracoid process, projection,	1	4½	8
12. Ditto, breadth at middle,	...	3	3½
13. Acromion process, anterior border,	...	7	8
14. ,, ,, posterior border,	...	3	4½
15. Thickness of scapula at middle of upper border,	1¼	0¼	0¼
Weight of scapula, in ounces, right,	558½	286	...
,, ,, left,	528	255	480¼

strongly and nearly uniformly arched form of the upper margin would be diminished in the more mature state, judging by the outline of the cartilage shown in the figures of Eschricht (xvii. p. 79) and that of Van Beneden and Gervais (Pl. X. and XI. fig. 6). These imply that the posterior angle would be elongated, and the upper border near it filled up and more flattened by farther ossification. The figure of Rudolphi (Taf. i. fig. 1) shows this to some extent. But the cartilage in B. musculus would also render its border still more straight. In my 64-feet-long B. musculus there was but a narrow strip of cartilage along most of the border, enlarging into a triangular plate behind and before, covering the more curved parts of the posterior and anterior ends of the bone, and forming a nearly straight upper border, with a little curving down at the posterior angle, the cartilage at which was the largest. D'Alton's figure of the scapula of Megaptera (Taf. iv. fig. *f.*) is the likest to the scapula of this Megaptera of any of the figures given ; but in his figure the anterior border is longer than the posterior, and the anterior angle is blunted. On the anterior border in my Megaptera, below the junction of its lower and middle thirds, there is a gentle elevation, 3 to 4 inches long, with a rough summit, shown only in the figure of Megaptera Lalandii of Van Beneden and Gervais (Pl. IX. fig. 4), scarcely to be recognised in B. musculus.

The Upper Border.—When these scapulæ are placed in pairs on the floor, standing on their glenoid cavities, various differential characters come into view; the relation is seen of the different thicknesses and curvatures of the upper border to the stronger and thinner parts of the body and to the curvatures of the surfaces. The much greater thickness of the upper border and of the whole bone in Megaptera is striking. In the 60½-feet-long B. musculus the thick parts are, in front for 8 inches (increasing from ½ inch to 1 inch forwards), the front half of that much bent down ; and behind for 16 inches (increasing from ¾ inch to 1¼ inch backwards), the hinder three-fourths of that much bent down. These thick parts appear as if incompletely ossified, as when denuded of cartilage. The long (32 inches) intervening nearly straight part of the border is thin, from ½ to ¼ inch, mostly ¼. These thick fore and back parts are also somewhat bent outwards, giving a moderate outward concavity to the border where each of the two thick parts meets the intervening thin part, but the general effect is a slight concavity outwards of the whole border. The two thick parts are seen to be the ends of the anterior and posterior beams of the scapula, the anterior beam strengthened by the spine of the scapula running up to it at the anterior angle. The same thick parts and curvatures are seen in the

50-feet-long B. musculus and in B. borealis, but in the latter
the anterior and posterior angles are bent rather inwards.

In *Megaptera* the whole upper border is of great thickness,
except for about 9 inches just in front of the middle, where it
is of but moderate thickness. The thickness of the hinder half
(27 inches in length) increases backwards from $1\frac{3}{4}$ to nearly
2 inches. The anterior thick part (18 inches in length) increases in thickness forwards from 1 to $1\frac{1}{2}$ inches. The thinnest
part, above defined, is $\frac{3}{4}$ inch thick. The posterior half has a
marked general curve, concavity inwards, the bay 1 inch deep.
The anterior half is gently sigmoid, the thinner part concave
inwards, the anterior and greater part convex inwards. Thus
the anterior angle is bent outwards as in B. musculus, while the
posterior angle is bent inwards. There are thus on the inner
side of the border two well-marked concavities, one on the
posterior half, the other just in front of the middle ; but on the
outer edge the convexity corresponding to the anterior concavity is very little marked, owing to the thinning on the thin
part being on the inner side. The beams of the scapula, reaching up to the thick parts on the border, are thus seen to be
broader than in B. musculus, the only thin part of the whole
scapula being that ascending from before the middle of the
glenoid cavity to the thin part of the border, showing itself as
a hollow in both directions on the outer surface of the bone.

In Megaptera, the *thickness of the bone* increases on about
its upper third, so that the border, all along, is thicker than
the part of the bone near it. But in B. musculus, the thickness
continues to diminish upwards to the border, all along, except
where the border is very thick, close to its posterior and
anterior ends.

Viewing now the *surfaces* of the scapula, the inner (venter)
presents none of those sharp ridges, radiating from the neck, which
are so well marked in *B. musculus*. In it they are seven or eight
in number, two running from the anterior border, the others
radiating from the neck, better marked on the anterior than on
the posterior half of the venter, and fully as well marked in the
50-feet-long specimen as in the more mature ones. These
ridges, with their intervening fossæ, give the whole surface of
the venter a fan-like appearance. Viewed as a whole the

venter in B. musculus has, apart from what is given by the
rise towards the glenoid margin, very little concavity. Traced
upwards from the neck, it is at first concave, then convex.
That convexity is very strongly marked in my 64-feet-long
specimen, so as to give a second concavity towards the top.
Viewed longitudinally, the venter is, on the whole, convex,
owing to the bending outwards of the angles, but, for about
the middle half or more, between the anterior and posterior
beams, there is some general concavity, about $\frac{1}{2}$ inch deep, with
depths of about 1 inch at the fossæ between the ridges. A
general concavity of the venter is better marked in B. borealis,
to a depth of $1\frac{1}{4}$ inch, and it has the same radiating ridges as
B. musculus, though they are less prominent. In *Megaptera*
the venter, traced vertically, is, above where it is influenced by
the neck, almost flat except along the middle part where the
bend outwards and broadening at the top causes a slight vertical concavity. Traced longitudinally, the posterior beam,
occupying the posterior half of the bone, shows a shallow concavity along its middle following the great convexity here on
the dorsum. Another shallow concavity runs up between the
two beams, at about the junction of the anterior and middle
thirds of the bone, corresponding to the thin part of the upper
border and to part of the great concavity on the dorsum. But
the whole ventral surface strikes the eye as flat and smooth
compared with that of B. musculus. Taken from the glenoid
margin, the venter has a depth, in Megaptera, of 2 inches; in
the 50-feet-long B. musculus, $\frac{1}{4}$ inch less; in the $60\frac{1}{2}$-feet-long
one, $\frac{1}{4}$ inch more than in Megaptera. The absence of the
radiating ridges for the intermuscular septa of the subscapu-
laris muscle, would seem to indicate a less development of that
muscle in Megaptera. The greater thickness of the bone has as
it were filled up the radiating fossæ between the ridges, but the
ridges were no less necessary for the fibrous septa which so
much increase the origin of the muscle.

The *dorsal surface* in Megaptera is more simple. Traced
longitudinally, it is convex at each beam, the posterior convexity
occupying about half, the anterior about a fourth, of the whole
surface, with a marked concavity between, corresponding to the
thin part of the upper border. A line between the most

prominent part of the two convexities gives a bay nearly 1 inch in depth. Traced vertically, there is very little concavity on the upper part of the surface, except on the posterior third. Taken from the glenoid margin, the concavity of the dorsal surface is $3\frac{1}{2}$ inches deep at the great hollow between the beams, $2\frac{1}{2}$ inches deep at other parts. The greatest depth of these fossæ is at about the junction of the lower and middle thirds of the bone. In *B. musculus*, the characters on the dorsum also are different. Traced antero-posteriorly, there are the convexities of the anterior and posterior beams, between these a wide general concavity, about 1 inch deep, intersected by a vertical ridge at about the middle of the bone, corresponding to one of the deepest radiating fossæ of the venter. Traced vertically, the upper part of the surface is a little concave on the anterior third, convex on the posterior third. Taken from the glenoid margin, the depth of the concavity of the dorsal surface is, high up on the anterior third, $2\frac{1}{4}$ inches, on the rest, about $1\frac{1}{2}$.

The *glenoid cavity* is not only longer than in the mature B. musculus, but is broader in proportion to its length, as seen by the measurements given in Table II. The greatest diameter of the articular head of the humerus in Megaptera is $12\frac{1}{4}$ inches, that of the $60\frac{1}{2}$-feet-long B. musculus is $9\frac{1}{2}$ inches. The greater breadth of the glenoid cavity in Megaptera, as regards outline form, is gained on both sides, but the neck and cavity project more to the outer than to the inner side, to the extent that $\frac{2}{3}$ of the breadth of the cavity lies to the outside of the plane of the bone. This is not the case in B. musculus, in which the projection to the two sides is either equal or greater to the inner than to the outer side. This difference may be in part owing to the presence of a large coracoid in B. musculus, and to the absence of an acromion process in Megaptera, but it is seen further back and is well marked.

Weight of the Scapula.—The greater robustness of the scapula of Megaptera is shown also by the weights given at the foot of Table II. In the 40-feet-long Megaptera it weighed 528 oz., against 255 oz. in the 50-feet-long B. musculus, and against $480\frac{1}{4}$ in the $60\frac{1}{2}$-feet-long B. musculus. The difference between these two latter is also striking. The right is heavier than the left in Megaptera by 30½ oz., and in the measurement (that given in the Table) is 1 inch higher and 1 inch longer than the left. The right in the 50-feet-long B. musculus is 31 oz. heavier than the left, and is ½ inch higher and ½ inch longer than the left. The left in the B. borealis is the heavier, by 141 oz. against 135 for the right; but the left has an additional piece behind, giving it a length of 30 inches against 28½ for the right.

5. THE HUMERUS.—Viewed from the shoulder, the articular surface of the head of the humerus seems more extensive relatively than in the other finners. This is owing to the articular surface in Megaptera advancing more than in them towards the flexor aspect. Hence the head in Megaptera is placed more on the end of the bone, although its projection to the exterior aspect, and olecranonwards, is not less than in other finners. The epiphysis of the head forms nearly a third of the length of the bone.

Where the articular cartilage of the head, and that of the tuberosity, had lain, the rough appearance of the bone is very striking; perforated by numerous vascular apertures, admitting a crow-quill or thick pin, the apertures surrounded by elevations, forming irregular prominences as thick as the end of a little finger, and joining each other in a network, rendering the whole area tubercular, pitted, and perforated.

6. THE FORE-ARM.—Besides by their greater length and thickness, as compared with those of this 50-feet-long B. musculus, as seen in Table I., the bones of the fore-arm in Megaptera may be recognised by marked differences in form. This is more strikingly seen by comparison with the paddle of my full-grown B. musculus (this *Journal*, 1872), 65 to 66 feet long, the paddle from head of humerus to tip 7 feet 8 inches, in which the radius, omitting the epiphyses, has precisely the same length as in this 40-feet-long Megaptera with a 12-feet-long paddle.[1]

The *radius* of Megaptera is less bent, is narrower along the

[1] In connection with the description of the bones and joints reference may be made to figure 6, showing the left paddle, flexor aspect, of this Megaptera reduced to $\frac{1}{14}$, in which I have endeavoured to represent all the parts with exactness; and, for comparison, to figure 13, showing the same aspect of the left paddle of the 64-feet-long B. musculus, reduced to $\frac{1}{18}$, also drawn by me from nature.

proximal third, expands greatly in **breadth towards** its carpal
end, and is thicker throughout.   At the **carpal end** the breadths
are 11¾ inches in Megaptera, in **this** full-grown B. musculus
only 8⅔.   This expansion in Megaptera renders the lower border
of the radius very concave along **its distal** half or third.   In the
50-feet-long B. musculus there is very little concavity **there;**
in **my 64-feet-long** B. musculus (this *Journal*, 1871), **the**
concavity **is more** marked, and most so just below the middle
of the bone, **but is not to be compared to the great concavity**
towards **the wrist in Megaptera.**   In the 65 to 66-feet-long
B. musculus **the radius is** convex throughout on this border,
except just after the proximal end.[1]

Ulna.—The differential characters of the ulna **in** Megaptera
are, the much **shorter and much less expanded** olecranon, its
shortness at the carpal end compared **with** the radius, the more
bent shaft, **and the greater** expansion at **the carpal end.**   The
breadths at the wrist are in Megaptera 7½ inches, in the 65 to 66-
feet-long B. musculus 6½.   The ulna falls short of the radius at
the wrist, **in** Megaptera by 3 inches, in this B. **musculus by**
1¼ **inches.**   The *bony* olecranon in Megaptera is a short blunt
process projecting 1 to 1½ inches beyond the **humerus, with an**
abrupt oval end, 3 inches in length, 2½ in thickness.   **In this**
65 to 66-feet-long B. musculus, it projects 5 inches beyond the
humerus and expands to a breadth of 12½ inches.   **In the 50-**
feet-long B. musculus, the bony olecranon projects 2½ inches,
and expands to a breadth of 7 inches, the future recurved **part**
represented as yet only by cartilage.   **In the 65 to** 66-feet-long
B. musculus this great bony olecranon reaches for 5 inches
along **the** ulnar border of the humerus, forming a nearly
rectangular socket for this part of the elbow-joint ; in Megaptera **not at** all, leaving **the here recurved** epiphysis of the

---

[1] By the above characters I am able to recognise a large cetacean radius which
came to Leith some years ago in a cargo of guano, as that of a full-grown
Megaptera.   Length, epiphyses consolidated, 37 inches ; breadth at wrist 12,
probably 13 inches, as it is somewhat injured here ; thickness at wrist, 7 inches.
On section it is seen that there is no medullary canal, but cancellous tissue
throughout.   The cancelli are more open at the second quarter of the bone than
at any other part.   There is a layer of dense tissue at the radial end, ¼ to ⅓
inch thick ; and along the middle half of the shaft the tissue may be termed
dense for ½ to ¾ inch at the surface, but not quite dense, and with gradual
transition to the cancellated part.

humerus exposed for 3 inches. The *cartilaginous* olecranon in the 65 to 66-feet-long B. musculus is 7 inches in height, 9 inches in length at the top. In the 50-feet-long B. musculus, the height is 6 inches along the middle, the length at the top 7 inches. The dimensions of the cartilaginous olecranon in this Megaptera were not noted when it was moist, but it was much shorter and narrower than in B. musculus, and now, in the dried condition, it is 4 inches in height and the same in length, and has probably shrunk about 1 inch in both directions.

The *interosseous space* of the fore-arm is much narrower in the Megaptera than in B. musculus. At the middle, the width in Megaptera is $\frac{3}{4}$ of an inch; in the 50-feet-long B. musculus fully 2 inches; in the 65 to 66-feet-long B. musculus $\frac{1}{4}$ inch less, this probably owing to the greater thickness of the bones. In Megaptera it is rather wider ($1\frac{1}{4}$ inch) just below the heads of the radius and ulna, and thereafter remains pretty equable at $\frac{3}{4}$ inch, lessening to $\frac{1}{2}$ inch just before the epiphysis of the ulna. In B. musculus it narrows a little towards the elbow, and beyond the middle diminishes gradually by the expansion of the radius, and ceases a few inches from the carpus, the borders becoming flattened for contact of the bones. This flattening is 5 inches in length in the 65 to 66-feet-long B. musculus, 2 inches in the 50-feet-long B. musculus, but the actual contact appears to have been for about half of these lengths. In this Megaptera there is no flattening of the interosseous borders and no contact of the radius and ulna, the shaft of the radius resting on its ulnar side on the forward-projecting part of the ulnar carpal bone. This narrowness of the interosseous space in Megaptera is in part due to the less concavity of the radius, but mainly to the greater curvature of the ulna. The curve of the ulna is on both its borders. A line drawn from the ulnar edge of the humerus to the upper edge of the ulna at the wrist gives in Megaptera a bay 4 inches deep; in the 65 to 66-feet-long B. musculus, 3 inches; in the 50-feet-long B. musculus, $2\frac{1}{2}$.

The fore-arm of Megaptera is still more differentiated from that of Mysticetus, in which the bones are comparatively short, the radius broad and flat, and the interosseous space wide.

7. THE ELBOW-JOINT.—The elbow-joint was diarthrodial, with one large synovial cavity. The cavity was continued on the

olecranon for $2\frac{1}{2}$ inches ($1\frac{3}{4}$ of it on the cartilaginous olecranon), with a partial deficiency across the sigmoid cavity, as in man; and was also continued in between the epiphysis of the radius and ulna for $1\frac{1}{2}$ inches. The facet of the humerus for the radius is almost flat; that for the ulna concave both ways, the hollow $\frac{1}{2}$ inch deep, till near the olecranon part where it becomes convex. The synovial membrane of the elbow passed over the edges of the bones for $\frac{1}{4}$ inch, until it met the ligaments. The elbow-joint allowed of a very little gliding motion.

8. THE CARPUS.—The carpus has six cartilages, representing bones, four in the proximal row, the radiale, intermedium, ulnare, and pisiforme; two in the distal row, in series with digits III. and IV. (see figures 6 and 9). Only the radiale and ulnare have ossification, seen only in section. The position and relative size of these small ossifications are shown in figure 9, as are also the ossification within the epiphysis of the radius, and the very small ossification within the epiphysis of the ulna. The cartilages representing the future carpal bones are well marked out on both the flexor and extensor aspect by surface grooves, bridged over by fibrous tissue, and the separation of the cartilaginous blocks goes through and through, the narrow interval occupied by soft connecting fibrous tissue, resisting the handle of the knife and allowing of a little motion. This is beautifully seen on horizontal section of the entire carpus, the great blocks of cartilage mapped out sharply as in an outline diagram. The numerous vascular perforations are seen over the whole area of the cartilages. The lines of articulation, occupied by the soft fibrous tissue, vary in width from $\frac{1}{16}$ inch to half that or less. The same kind of articulation, by intervening fibrous tissue, is seen at the wrist-joint, but with a somewhat wider interval (about $\frac{1}{12}$ inch), and at the carpo-metacarpal articulations.

Synovial cavities exist at both the proximal and distal ends of the cartilage of the second row (os magnum) which supports digit III. The proximal of these cavities reaches across the whole breadth of the end (2 inches). The distal cavity covers only about a fourth (1 inch) of the distal end, where the cartilage forms a blunt peak. Both of these synovial cavities go through and through from flexor to extensor aspect. It will be observed that these diarthrodial joints occur on the ends of that carpal cartilage which supports the great digit.

On section, the separation of the ulnare and intermedium is less striking along the proximal half of their articulation, but it exists, and there are the usual surface grooves on both aspects. No separation is visible between the pisiform and the epiphysial cartilage of the ulna along the proximal $\frac{2}{3}$ of their relation, but the surface grooves are complete. A faint groove, or break, appeared on the surface seeming to subdivide the very broad ulnare into an ulnar and a radial portion, but as on section and on slicing near the surface no trace of separation could be seen, it was probably unnatural.

Comparison with the Carpal Bones of other Whales.—As I hope to go into this subject fully in a subsequent paper, a short statement may here suffice. Megaptera differs from the other finners, and from Mysticetus, in the enormous extent of the ulnare (cuneiform bone), reaching as it does to opposite the ulnar fourth of the carpal end of the radius, and expanding to meet the short ulna. It is thus broader (10 inches) than the intermedium (semilunar bone) and radiale (scaphoid bone) together (about 4 inches each), and is also, from fore-arm to metacarpus, the longest carpal cartilage. The ulnare in the other finners does not reach to opposite the radius. The two cartilages of the distal row (os magnum and unciform) are much in the same position as in the other finners, but in B. musculus the trapezoid bone, or cartilage, is present, though not in B. borealis. The separation of the pisiform cartilage from the epiphysis of the ulna is complete in B. borealis. In the 50-feet-long B. musculus it is not so marked in the proximal as in the distal half of the articulation, but it exists. I would not regard the incomplete separation of the pisiform cartilage as a character of Megaptera. It is the same in my half-grown B. rostrata.

The epiphysis of metacarpal V. in Megaptera might readily be taken for a carpal cartilage, and when the bony metacarpals are removed so might the epiphysis of metacarpal II. It is notable that though digit II. is so much more massive in Megaptera than in other finners, there is no trapezoid, much less a trapezium, bone present in it.

In both the 50-feet-long B. musculus and the B. borealis,
the carpal bones are well ossified, especially those of the first
row, except the pisiform which remains entirely cartilaginous in
all of them. This contrasts with the little progress ossification
has made in the carpus of this 40-feet-long Megaptera.[1]

9. THE CARTILAGES AND JOINTS OF THE DIGITS.—The cartilaginous enlargements between the digital bones are greatly
developed in Megaptera. Each of these nodes is composed of
the two epiphysial cartilages, separated by a more or less
developed synovial cavity.[2]

Mode of Articulation.—The perichondrium enveloping the
node is, like the periosteum, about ⅔ inch thick, and, without
removing this membrane, a transverse depression is seen indicating the presence of a joint. In the earlier nodes the position

---

[1] *Vascular Canals of the Cartilages in Cetacea.*—It may be noted here that these
and all masses of cartilage in cetacea are pervaded by channels for red blood-
vessels, resembling the system of Haversian canals in bone. On separating the
epiphysial *cartilages of the digits* from their phalanx, the contained blood-vessels
are seen like a system of cordage passing from the bone into the cartilage, stretching for about an inch before they give way. In vertical transverse sections they
are seen as rounded apertures, with occasional communications, the apertures
from ⅛ to ¼ inch apart, giving a perforated appearance to the whole area. In
horizontal sections they are seen as a system of longitudinal canals, with anas-
tomoses. In vertical longitudinal sections (flexor to extensor aspect) they appear
fewer in number, some transverse but mainly longitudinal; seemingly none
vertical, as if the vessels did not enter from without, but on both surfaces
are seen a moderate number of apertures of considerable size. The channels
advance from the bone to the joint and are seen from the synovial surface as
dimples with a thin covering. These may readily be converted into and mistaken for apertures. They give the synovial area a mottled appearance. In
horizontal section through the middle of the *carpus*, the blood-channels are seen
throughout the great masses of cartilage, at distances of ⅛ to ¼ inch apart, some
cut transversely, most cut obliquely, as if going in from the surface, with anas-
tomoses. On vertical section, not many are seen going in from the surface, but
on the surface itself a number of large apertures are seen. The blood-vessels of
the carpal cartilages must come from the surface, and the blood-channels within
the cartilages seem to form a system striking in vertically but forming an anas-
tomosing network.

[2] We make much of the fact of the epiphysis in man and various mammals
being distal on the metacarpal bones and proximal on the phalanges. I called
attention in 1863 (*Edin. New Phil. Jour.*, July 1863) to the fact, with which
I had been long familiar, that in the cetacea there is an epiphysis at both ends of
each phalanx and also at both ends of each metacarpal bone. This epiphysis at
both ends of all these bones may be readily seen in an ossified condition in the
paddle of Globicephalus melas and of the common porpoise. But I have not
found it in an ossified condition in any whalebone whale, and Megaptera is no
exception to this.

of the cavity is proximal to the middle of the node; this is very marked in both the nodes of the index digit, while in the more distal nodes it is rather distal to the middle of the node. On dividing the perichondrium the cavity opens freely, exposing the two smooth synovial surfaces, and extends over the whole of the ends, except in the three distal nodes of the two great digits (III. and IV.) and the distal node of digit V., in which it does not reach quite to the extensor surface or fully to the sides, especially not on the side to which the node projects most. But the depression indicating the line of the joint is seen equally on the extensor surface. The part not separate is occupied by soft fibrous tissue, showing as a white line on section. As bearing on the incompleteness of diarthrodial articulation in the above-mentioned more distal nodes, it may be remarked that these nodes themselves are flexible, being flat and thin compared with the more proximal nodes, which are very thick.

Form of the Articular Surfaces.—The surfaces are generally slightly concavo-convex, the convexity generally on the proximal cartilage, but the curving is less in the direction between the flexor and extensor aspects than transversely in the direction from digit to digit. This departure from the flat in the interphalangeal joints is less in the more proximal nodes. At the metacarpo-phalangeal joints the surfaces are more nearly flat, or are sinuous, varying as follows:—The metacarpal is, on the index digit, slightly convex; on digit IV. slightly concave; on digits III. and V. it is sinuous, the concavity on the ulnar half in digit III., on the radial half in digit V. The metacarpophalangeal joints of digits II. and IV. are nearly opposite to each other in the limb.

The Digital Joints in Megaptera in comparison with those of other Whales.—The synovial cavities are well developed in Megaptera. In my 64-feet-long B. musculus the cavity, indicated on both surfaces, extended across only the middle third, or less, of the node, and was absent in the smaller nodes. In Mysticetus there are the surface furrows and there is motion between the two parts of the node, but the boundary, which is seen through and through, is not effected by a cavity but by a layer of very soft cartilage.

C

Even when synovial surfaces occur in cetacean digits, the motion can be only that of a little gliding on the nearly flat surfaces. The cartilaginous nodes of the digits provide, by the flexibility of the substance itself, for a certain amount of general bending along the distal part of the paddle.

The Terminal Cartilages of the Digits.—The terminal cartilages of digits III. and V. resemble each other in that there is a cartilage attached to the distal bone, and a joint across part of the cartilage. In *digit III.* the cartilage attached to the distal end of the 8th bone is horse-foot shaped in outline, $2\frac{5}{8}$ inches across by $1\frac{3}{4}$ longitudinally (fig. 10). The joint is situated $\frac{1}{2}$ inch from the tip, is $1\frac{1}{4}$ to $1\frac{1}{2}$ inch in breadth, reaching to $\frac{1}{2}$ inch from the radial border and nearer to the ulnar border, and goes through the whole thickness of the cartilage. It is curved, the concavity towards the bone. The joint is a soft part of the cartilage, and permits of motion. This joint had the same position in both right and left digit III.

In *digit V.* the cartilage, attached to the distal end of the 4th bone, is ovoid in outline, wanting the proximal fourth of the ovoid where it is fitted on the phalanx; is $1\frac{3}{8}$ inch across and $2\frac{3}{8}$ longitudinally. The joint is situated about $\frac{3}{4}$ inch from the junction with the phalanx, occupies the greater part of the breadth, is sinuous but mostly with its concavity towards the bone, and the cartilage bends at it readily. It was, however, seen only on the flexor surface, and, on section, to go only a third to half-way in from that surface. This was on the left paddle; in the right paddle no joint could be detected in the distal cartilage of digit V

In *digit II.* the cartilage, attached to the distal end of the 3rd bone, is very greatly expanded, is of broad horse-foot shape in outline, $4\frac{1}{2}$ inches across, 3 inches longitudinally (fig. 11). This cartilage forms most of the great hump on the radial border of the paddle, is about midway between the radial hump and the tip, and must give not only breadth but strength and resistance to this part of the paddle. It presents *two joints,* dividing it transversely and incompletely into three nearly equal parts, the proximal of the two the most marked. They are in part diarthrodial, in part formed by soft connective tissue. Both have the concavity proximal, parallel to the end of the bone and to the distal border of the cartilage. The proximal joint is a cavity for more than the middle half of the breadth of the cartilage, and for the rest is soft connective tissue. Towards the radial side, the line of the joint, in the left paddle, makes a bend with the concavity distal on the radial fourth of the joint; in the right paddle, it is bifurcated there like the letter ≺. The distal joint occupies about the middle half of the cartilage and presents a central synovial cavity. It was noticed only on section. The preserved sections show the elliptical synovial cavity of the distal joint concealed on both surfaces, and the proximal joint as a more developed synovial cavity, indicated by furrows on both the flexor and extensor surfaces, but the cavity may be interrupted in part by soft fibrous tissue (fig. 12). As this distal joint would have been overlooked had the surfaces only been examined, we see how the commencement of an additional phalanx, or the existence of a phalanx potentially, may be overlooked.

In digit IV., after the 7th node, there is a solid cartilage, 2¼ inches in length by 2 in breadth, which might pass for a repetition of the 7th bone, or of the 8th bone of digit III., but for greater enlargement at its distal end, on the radial side. It is plainly the 8th bone of the digit as yet unossified. Then, with a joint between, comes a narrower cartilage, 2 inches in length by 1 inch in breadth.

Interpreting these terminal cartilages and their joints, the cartilage and its dividing joint in digits III. and V. might be described simply as the parts of a node in a flattened form, with its rudimentary joint. If more phalanges are to be formed, then the part distal to the joint would be potentially a phalanx and its epiphysis, while the part proximal to the joint would be the distal epiphysis of the as yet distal phalanx. The simplest view is to regard the part between every two joints, or distal to the distal joint, as an element capable of becoming a bony phalanx with its unossified part, or epiphysis, at each end. In digit IV., the first mentioned cartilage represents a phalanx and its distal epiphysis; the distal cartilage, the same as the element distal to the joint in the terminal cartilage in III. and V. In digit II., in which the great expanded cartilage has two joints, the proximal part may be regarded as the distal epiphysis of the distal bone, and the parts on either side of the distal joint as potentially a phalanx and its two epiphyses. Possessing the cartilaginous elements above indicated, ossification might thus give additional phalanges as follows, to digit II., 2; to digits III. and V., 1; to digit IV., 2.

10. THE BONES OF THE DIGITS—*Relative Length of the Digits.* —Referring to Table III., it is seen that digit III. is the longest, as in B. musculus, but that digit IV. comes rather nearer to it in length than in B. musculus. Also that digits II. and V. are short relatively to digits III. and IV., being under half the length, while in B. musculus they attain to about two-thirds of the length of the two long digits. In Megaptera the radial digit II., the index, contrasts greatly with the ulnar digit V., in the robustness and length of its bones compared with those of B. musculus. With a phalanx less than digit V. it maintains its greater length; its three bones are not far from twice the length of the corresponding bones of digit V., and are several times their thickness, while in B. musculus the difference in length and in the thickness of the bones of these two digits is not great. The enormous robustness in particular of the metacarpal bone of the radial digit in Megaptera almost reminds one of the metatarsal of the human hallux, having the same adaptation, giving resistance to that side of the limb.

Number of the Digital Bones.—The number of the bones, including the metacarpal, in each of the four digits, is,—digit II., 3; digit III., 8; digit IV., 7 (and a cartilage); digit V., 4. Beyond these, each digit has the terminal cartilage, those of digits V. and III. with a joint in them, that of digit II. with two successive joints in it.

The following Table, giving the number of ossified bones, including the metacarpal, found in each finger in my four dissections of B. musculus (all males), shows how the number may vary according to maturity or otherwise. The Table does not give the terminal cartilages, which might subsequently have become ossified. The number in the B. borealis (male), also given in the Table, is the same as in the 50-feet-long B. musculus.

11. TABLE III.—*Number of Ossified Bones in each Digit.*

	II.	III.	IV.	V.
Balænoptera musculus, 65–66 feet long, . . .	5	8	8	5
,,　　　　,,　　60½ feet long, . . .	5	7	6	4
,,　　　　,,　　64 feet long, . . .	4	7	6	4
,,　　　　,,　　50 feet long, . . .	4	7	6	4
Balænoptera borealis, 35 feet long, ...	4	7	6	4
Megaptera longimana, 40 feet long,[1] . . .	3	8	7	4

In the 60½-feet-long B. musculus, digit III. is somewhat deformed, the three distal bones detached, narrow, and partly pushed up between digits III. and IV. The 5th bone of its digit II. is a very narrow cone about an inch in length. This whale was mature if not aged. Its pisiform cartilage is partly ossified. In the 65 to 66-feet-long B. musculus, which may be regarded as aged, the 5th bone of digit II. is 2¼ inches in length; the 5th bone (now lost) of digit V. was 2½ inches in length. The 8th bone of digit IV. is small, ⅝ inch in length. I have no note as to the presence of a terminal cartilage in addition to the 5, 8, 8, 5 bones. In the 64-feet-long B. musculus each digit had a terminal cartilage, in addition to the 4, 7, 6, 4 bones (this *Journal*, 1871).

Looking to the state of the terminal cartilages above described, it would appear that this Megaptera has unossified cartilages in

[1] The number of bones in each digit of Megaptera given by Van Beneden and Gervais (*Ostéographie des Cétacés*, completed 1877, Pl. X. and XI., fig. 5, the Megaptera longimana of the Brussels Museum; Text, p. 129) is, including the metacarpal, 3, 8, 8, 4; beyond these a terminal cartilage is figured, though not of the same shape as in my Megaptera. The number figured by D'Alton, in his fasciculus (*Die Skelete der Cetaceen*, 1827, Taf. III., fig. e), is 3, 8, 7, 3, with what appears to be intended for a terminal cartilage on each. The form of the phalanges is not well represented. In the small figure given by Rudolphi of the skeleton of his 43-feet-long Megaptera longimana (*Loc. cit.*, Taf. I., fig. 1) the numbers shown, including the metacarpals, are 3, 9, 7, 4; but he mentions that the 9th of digit III. "hat keine Knochensubstanz." The number given by Eschricht (*Loc. cit.*, p. 141) is 3, 9, 9, 3. His figures are from two fœtuses. In that from his 45-inch-long fœtus (Taf. III., fig. 4) his enumeration, 3, 8, 8, 3, is in addition to the metacarpal bones. The terminal pieces in the figure would correspond numerically to what I have described above as the terminal cartilage, except in his digit V. which would require an additional piece. In his figure (XVIII. p. 79) of the 35-inch-long fœtus, if the metacarpal and the terminal cartilage are included, the numbers are, 4, 10, 10, 4. In my Megaptera the inclusion of the metacarpal and the terminal cartilage would make the numbers 4, 9, 9, 5; or in view of the two joints in the terminal cartilage of the index, 5, 9, 9, 5.

digits II. and V., which bring their number up to that of the ossified bones in the mature B. musculus; and unossified cartilages in digits III. and IV., which would, if ossified, bring their number up to 9 each. Megaptera, therefore, has at most, if at all, one more phalanx in its two long digits than B. musculus. The great length of the second digit in *Globicephalus melas* (Digit I., 4 bones; II., 14; III., 9; IV., 3; V., 1) is obtained by increasing the number of the bones, but Megaptera, true to its affinities, gains the great length of its two long digits by elongation of the bones.

12. Table IV., *showing, in inches, the Length of the Digits and the Length, Breadth and Thickness of the Bones in each. Farther Cartilages are indicated by the letter* C.

Whale.	Megaptera longimana.				B. musculus, 50 feet long.[1]			
Digit.	II.	III.	IV.	V.	II.	III.	IV.	V.
Total length from carpus to tip of distal cartilage, .	38	81	80	32½	17¼	25½	23	15½
1st bone (Metacarpal), length,	9	12	9	5¼	3¾	5	4⅛	3¼
,, ,, breadth,	4¼	2¾	2	2¼	1½	1⅜	1¼	1⅜
,, ,, thickness,	3⅛	2⅛	1⅛	1¼	1¼	1¼	⅞	⅞
2nd bone ,, length,	10	11¼	9¼	5⅜	3⅝	4½	4	3
,, ,, breadth,	2⅛	2⅓	1⅝	1⅛	⅞	1¼	1⅛	1
,, ,, thickness,	2¼	2½	1½	⅞	⅝	⅞	⅝	½
3rd bone ,, length,	7¼	9½	9	4¾	3¼	3¾	3⅛	2⅜
,, ,, breadth,	1½	2⅜	1½	1	⅝	1⅝	1	⅜
,, ,, thickness,	1⅕	2¼	1⅛	⅓	½	⅞	½	¼
4th bone ,, length,	C	7½	7⅝	2¼	2	2¼	2⅜	1¾
,, ,, breadth,	...	2⅓	1½	¾	6/16	1⅛	⅝	⅞
,, ,, thickness,	...	1¾	1¼	⅜	¼	½	⅜	¼
5th bone ,, length,	C	6	5⅝	C	C?	1½	1¾	C?
,, ,, breadth,	...	2¼	1⅝	...	...	1	⅝	...
,, ,, thickness,	...	1⅓	1	...	...	⅜	⅜	...
6th bone ,, length,	...	4¼	3½	...	...	1¼	1	...
,, ,, breadth,	...	1⅝	1¼	...	...	⅝	⅜	...
,, ,, thickness,	...	1⅛	¾	...	...	...	⅛	...
7th bone ,, length,	...	3	2¼	...	...	×	C?	...
,, ,, breadth,	...	1½	1⅕	...	...	...	...	...
,, ,, thickness,	...	⅞	½	...	...	...	...	...
8th bone ,, length,	...	1⅜	C	...	...	C?	C?	...
,, ,, breadth,	...	1⅝	...	...	...	...	...	...
,, ,, thickness,	...	½	...	...	...	...	...	...
9th segment (cartilage), .	..	C	C	...	...	...	...	...
The bones together, length, .	26¼	55⅛	46⅛	17⅞	12¾	17⅞	16⅝	10½
Leaving for the cartilages, .	11¾	25⅞	33⅞	14⅝	4½	7⅝	6⅜	4⅝
Proportion of cartilage to the entire digit, per cent., .	38	31·9	42·3	45	27·5	29·9	28·8	31·1

[1] In this B. musculus the state of the terminal cartilages could not be ascer-

Relative Length of the Digital Bones.—Running the eye down the columns of Table IV. it is seen that, in B. musculus, the lengths diminish gradually from the metacarpal onwards in the two long digits. But in Megaptera the first phalanx is longer than the metacarpal in all the digits except digit III., although its first phalanx is the longest of the first phalanges. This is owing to the still greater excess in length of the metacarpal of digit III., the metacarpal and first phalanx of which together exceed in length by $4\frac{9}{8}$ inches the two corresponding bones of digit IV., although the total length of digit III. exceeds that of digit IV by only one inch. In digit II. of the 50-feet-long B. musculus the first phalanx is slightly longer than the metacarpal, and in my more mature specimens of B. musculus it is so in the case of both digits II. and V., in the 65 to 66-feet-long specimen to the extent of from $\frac{1}{2}$ to $\frac{3}{4}$ inch, but they present no other exception to the progressive diminution in length onwards. The exceptionally great length of the metacarpal and first phalanx of digit III. in Megaptera, will be referred to in connection with the adaptation of the nodes and hollows of neighbouring digits.

Form of the Digital Bones.—In my more mature specimens of *B. musculus* the ends of the bones are concave towards the nodes, especially from side to side, most marked at the first two nodes of digit III. and the first node of digit IV., but most marked at the first node of digit III. On the lateral digits the bones are rather cut obliquely, so as to give a somewhat wedge-shaped node, the base of the wedge towards the free margin.

tained. The 7th bone of its digit III. was injured. In Table III. it is seen that in the most mature B. musculus digits II. and V. had 5 bones each, digits III. and IV. 8 bones each. In this Table, IV., the lengths are taken along the middle ; the breadth and thickness at the middle of the shafts, where the bones are narrowest. The breadth is from radial to ulnar, the thickness from flexor to extensor aspect. To show the much greater breadth at the ends of each bone than at the middle, I subjoin the following measurements, in inches, of the breadths of each bone of digit III. of Megaptera, at its distal and proximal ends, that of the proximal end placed first :—1st bone, $4\frac{3}{8}$, $4\frac{2}{3}$; 2nd bone, $4\frac{1}{2}$, 5 ; 3rd bone, $5\frac{1}{3}$, $5\frac{1}{16}$; 4th bone, $4\frac{3}{8}$, $4\frac{3}{8}$; 5th bone, $3\frac{3}{4}$, $4\frac{1}{16}$; 6th bone, $3\frac{1}{16}$, $3\frac{1}{4}$; 7th bone, $2\frac{1}{4}$, $2\frac{5}{8}$; 8th bone, $1\frac{1}{2}$, $1\frac{3}{4}$. It is seen from these measurements that, with the exception of the metacarpal, and slightly of the second phalanx, the expansion is greater at the distal than at the proximal end. This is seen, and to a more marked degree, on all the bones of digit IV. It does not apply, however, to the 1st and 3rd bones of digit II., or to the 1st and 2nd bones of digit V.

The measurements of these bones were all made along the middle to avoid fallacy arising from these variously shaped ends. In the *Megaptera* there is very little of the cupping at the ends, though it is discernible at the first nodes of digits II., III., and IV. But the oblique cutting of the ends in digits II. and V., giving the wedge-shaped node, is distinct, and the same is seen on digits III. and IV. after they have passed beyond the lateral digits, the base of the wedge towards the free border of each. This wedge form of the node must tend to give the digits more lateral movement from the axis of the limb, as in spreading the fingers.

In all the specimens of *B. musculus* the finger bones have the hour-glass form. The expansion at the ends is less on the metacarpals, owing to the greater breadth of the shafts, and is least at the carpal end of the two middle digits. The hour-glass form is less marked on the three distal phalanges of the long fingers and on the two distal of the shorter fingers, as the phalanges become more flattened. Roughly speaking, the expanded ends are about twice the breadth of the shaft, at the middle. The above applies to *Megaptera* also with the following peculiarities. In digit II. the metacarpal presents less expansion owing to the great robustness of its shaft, but its other two bones have the most hour-glass form of all the finger bones. The excavation of their shaft accommodates the first and second nodes of digit III., but is no less marked on their free border, owing to the great expansion at their ends to support the large nodes which project on their free border. The excavation on the ulnar side of the first phalanx of digit II. to receive the first node of digit III., flattens the border so much as to give the shaft a prismatic form. A like excavation and flattening is seen on the much smaller corresponding bone in B. musculus. The bones of digit IV. appear slender in the series in Megaptera compared with B. musculus, owing not only to the great robustness of digit III. but likewise to that of digit II. But comparing simply the two long digits in Megaptera and in B. musculus, the relative slenderness of digit IV. in Megaptera is striking,

Breadth and Thickness of the Digital Bones.—The measurements given in Table IV., showing the breadth and thickness of each bone in Megaptera and in the 50-feet-long B. musculus, give interesting results, read across and read down each column. Reading across, it is seen that, in each of these whales, both the breadth and thickness of the metacarpal bones, and of the corresponding range of phalanges, diminish progressively from digit III. to digit V., with the single exception of the metacarpal of digit V. which is $\frac{1}{8}$ inch broader than that of digit IV. In B. musculus the metacarpal of digit II. is likewise $\frac{1}{8}$ inch broader than that of digit III., and they are of the same thickness. In Megaptera the metacarpal of digit II. is seen to be greatly more robust than that of digit III., exceeding it in breadth by a half, in thickness by about a fifth.

Reading down the columns, it is seen that, in *B. musculus*, both the breadth and the thickness diminish progressively onwards, the only exception being that in digit III. its 2nd and 3rd bones have the same thickness, and its 3rd and 4th bones the same breadth. In *Megaptera* the exceptions to progressive diminution distally are in digit III., that the 3rd bone is the broadest ($\frac{3}{4}$ inch broader than the 2nd, and 1 inch broader than the metacarpal) and that the 4th bone has the same breadth as the 2nd; in digit IV., that the 2nd and 3rd bones have the same thickness, and the 3rd and 4th bones the same breadth.

Comparing the breadth with the thickness, irrespective of size, reading across each range, it is seen *in B. musculus*, that, from radial to ulnar side of the paddle, there is a proportionate diminution of the thickness, giving a progressive flattening of form from digit II. to digit V.; less marked from digit II. to digit III., well-marked at digit IV., and very marked at digit V.; and it is seen to be more marked as we read down the columns. The only exception is on the metacarpal bone of digit III., the thickness of which is only $\frac{1}{8}$ inch less than the breadth, while it is $\frac{2}{8}$ less on the metacarpal of digit II. In *Megaptera* there is not the same progressive flattening from the radial to the ulnar side of the paddle. Reading across the range, the bones that are thickest in proportion to their breadth are the 2nd of digit II., the 1st of digit III., and the 4th and 5th of digit IV. The bones of digit V have a more flattened form than the others in the range, but their length compared with their breadth renders this less striking to the eye.

Reading down the columns, comparing the breadth with the thickness of each bone, the flattening goes on progressively, *in B. musculus*, in all, becoming more and more marked. It is but little on the metacarpal bone of digit III., and on digit II. after the metacarpal. In my 65 to 66-feet-long B. musculus, as in Megaptera, the thickness of the 2nd bone of digit II. is greater than the breadth, the shaft being excavated to make room for the first node of digit III., but no other bone in digit II. of that B. musculus is thicker than it is broad. In *Megaptera*, the 2nd bone of digit II. is thicker than broad, owing to the adaptation above mentioned, and these two measurements are equal in the 2nd bone of digit III., and in the 3rd bone of digit IV. The metacarpals of digits III. and IV. in B. musculus are a little thicker in proportion to their breadth than the first phalanx is, which is not the case in Megaptera. Comparing digits III. and IV. of Megaptera, the first two bones of digit III. are thicker in proportion to their breadth than those of digit IV., but this proportion is reversed on the 3rd, 4th, 5th, and 6th bones of these two digits. Hence the more slender appearance of the bones of digit IV. in proportion to their length, viewed along the surface.

Viewing the digital bones of *Megaptera*, these measurements show that there is a variable proportion between the robustness and the length. In digit II. the metacarpal has the same length as that of digit IV but it is twice as robust. Its 2nd bone also, and to some extent its 3rd, are more robust, in proportion to their length, than the corresponding bones of digit IV. All along digits III. and IV., the greater robustness of the bones of the former, in proportion to their length, is striking. In digit III. the third bone is actually more robust than the 2nd, and it is $1\frac{3}{4}$ inch shorter. Its remaining phalanges also are more robust than the 2nd bone, in proportion to their length. This is after digit III. has ceased to be splinted on its radial side by digit II. In like manner in digit IV., the 2nd bone is not so robust, in proportion to its length, as the bones beyond it, but this is not to so marked an extent on the immediately succeeding bone as in digit III. In digit V., however, the 3rd and 4th bones are not more robust in proportion to their length than the 2nd bone.

13. PROPORTION OF CARTILAGE TO BONE IN THE FINGERS.—
The extent to which each finger is formed of bone and of cartilage
in these two whales is seen in Table IV. In the B. musculus
(50-feet-long) the proportions are to be taken as approximative
only, as in a more mature state the distal cartilages would be
in part ossified. But the proportions in my 64-feet-long B.
musculus correspond pretty well, being, of bone and cartilage,
respectively, in digit II., 18⅝ and 6⅛ inches; digit III., 25 and
8½; digit IV., 22½ and 8½; and in digit V., 14 and 5. The
comparison is most reliable on the two long digits. In the
Megaptera, the great terminal cartilage gives digit II. a large
percentage of cartilage, but in all its digits the percentage of
cartilage is greater than in B. musculus. The high percentage
of cartilage in digit IV. of Megaptera is to some extent gained
by the 8th element not being ossified, but is mainly owing to
the relative shortness of the more distal bones of the digit.
It would appear, therefore, that the great length of the fingers
in Megaptera, compared with those of B. musculus, while
mainly obtained by bone, is in still larger proportion gained by
cartilage. This should allow of greater flexibility, but may be
regarded rather as related to the greater robustness of the bones
in Megaptera. The large amount of cartilage in the fingers
of the cetacea, reaching in these two whales to about from $\frac{3}{10}$
to $\frac{4}{10}$ of the whole length of the digit, may be regarded as an
adaptation to general flexibility in digits so ensheathed that
their constituent bones cannot be moved separately.

14. ADAPTATIONS OF THE NODES AND PHALANGES TO EACH
OTHER IN NEIGHBOURING FINGERS.—The alternating nodes and
hollows fit into each other more closely than would appear
from the skeleton. The nodes are very large, most prominent
at the middle where they project ½ to ¾ inch on each side
beyond the level of the expanded ends of the bones. On the
surfaces, flexor and extensor, the nodes do not rise above the
level of the enlarged ends of the bones, except to form a gentle
convexity. The great enlargement is in the breadth. Stated
generally, the nodes are about three times as broad as the narrow
part of the phalanges at the middle of the shaft. The bones
and the nodes, both invested in their thick fibrous covering, form

a series of alternating great elevations and hollows, continued smoothly from one to the other, and mostly fitted to each other. The nodes are not generally opposite the middle of the hollows. That could only well be with equal length of the bones of neighbouring digits. The hollows are about twice the length of the nodes, and the node may be received into the proximal, or the middle, or the distal part of the hollow.

The alternation is accomplished simply thus. By the greater length of the metacarpal bone and also of the first phalanx of digit III. the nodes of that digit are projected just beyond the nodes of the two neighbouring digits, which, again, are about opposite each other. The result is that the nodes of digit III. are received into the proximal part of the hollow of the succeeding phalanx of the neighbouring digits, and that the nodes of the latter are received into the distal part of the hollow of the preceding bone of digit III. Then, by the shortness of the metacarpal of digit V., the adaptations between it and digit IV. are in like manner accomplished. If digits II. to V. of the human hand be placed together, it will be seen that there is the same arrangement of the phalanges in them, more evident if the observer will imagine the middle finger drawn forward a little. In these other fin-whales the method by which the nodes are rendered not opposite each other is the same, but the result is less striking in them than in Megaptera, owing to the comparative shortness of the bones and to the lesser size of the node compared with the breadth of the phalanges in them.

The following are the exact positions of the nodes in Megaptera. In digit III. their position opposite the proximal part of the hollow of the succeeding bone continues till the next last node, the 6th, is reached, which is seen to lie in the distal half of the hollow, owing to the shortness of the 6th bone of digit IV. The 5th node may be said to occupy the entire hollow of the 6th bone of digit IV. In digit IV., correspondingly, the fitting of the nodes against the distal part of the hollows of digit III., is continued on the first five; the 6th lies in the whole hollow, and the 7th more on the proximal side of the hollow. Of digit II. the first node lies in the distal third of the hollow of the metacarpal, the second just past the middle of the hollow of the 2nd bone of digit III. The expanded terminal cartilage of digit II. is opposite the proximal part of the hollow of the 2nd bone of digit III. but not close to it. The 2nd bone of digit II. receives the first node of digit III. at the proximal half, and is so excavated by it that the breadth of this shaft is less than its thickness. The 3rd bone of digit II. receives the 2nd node of digit III. at the distal half of its hollow. Digit V. has so short a metacarpal bone, and begins so much earlier at the carpus, that its first node lies even proximal to the middle of the metacarpal of digit IV. The second node is nearly opposite the first node of digit IV., its centre about an inch beyond the centre of the latter, but the node is very flat on the side next digit IV., and prominent on the free border. The 3rd node is opposite, but not close to, the distal part of the hollow of the 2nd bone of digit IV., and is very prominent on that side, but not on the free border. The hollow just beyond it corresponds to the 2nd node of digit IV. But digit V. is not in close relation with digit IV., though its metacarpal bone is not so divergent as that of digit V. is in the other finners.

Almost all the nodes which are on the free borders project more, and also rise to a sharper point, like mountain peaks, on the side next the free border. Exceptions to this were in the 3rd node of digit V. which, however, was $1\frac{1}{2}$ inch from contact, and in the 3rd node of digit III., the one succeeding the great terminal projection of digit II. For the most part the nodes in relation with neighbouring digits lie pretty close to them, the intervening spaces occupied by fatty and other soft tissues. Digits III. and IV. were only about $\frac{1}{4}$ inch from actual contact with each other at the 4th and 5th nodes of each. At other parts the interdigital space is wider, 1 to 2 or even $2\frac{3}{4}$ inches. Between digits II. and III. the space averages an inch, increasing distally. Digit V., after its 2nd node, is not near digit IV. At the tip, digits III. and IV are not close together, about $2\frac{1}{2}$ inches apart. Digit III. projects 2 inches beyond digit IV. though it is only 1 inch longer. It begins 1 inch later at the carpus than digit IV.

It would seem not improbable that the adaptation of these alternating nodes and hollows may, from this cause alone, to some extent determine the relative length of some of the phalanges. With digits near each other the nodes must lie in some part of the neighbouring hollows. That the joints should not be on a line with each other in the paddle is a farther reason for difference in the length of the phalanges. Had they been in a line with each other, transversely or obliquely, the paddle would have been liable to bend or break at the line.[1]

15. MUSCLES OF THE FINGERS AND FORE-ARM.—Considering the great size of the pectoral fin in Megaptera longimana, it was interesting to ascertain whether finger-muscles are present, and if present, whether they are more developed than in other finners, or still more rudimentary.[2] By incisions I was allowed to make in the fore-arm, when the Megaptera was on exhibition at Aberdeen in February 1884, I satisfied myself that red muscles were present, and I was able to dissect them fully in the autumn of that year. I found the same muscles present as in B. musculus, but instead of being larger, like the fingers on which they act, they were not half the size of those of

---

[1] It may be well to mention the means taken to secure accuracy in regard to these and other points in the anatomy of the paddle, besides the measurements. (1) An exact paper shape was cut of the entire paddle when attached to the carcase. (2) After the skin, fat, &c., had been removed in the dissecting-room, an outline of each digit was carefully traced on paper, showing the size and relations of the nodes and phalanges. (3) Before the bones were macerated, saw marks were placed on the proximal end of the flexor aspect of each, such as to enable us to articulate every bone in its right place.

[2] The presence of muscles in the fore-arm of a cetacean was first noticed by Professor Flower, C.B. (in B. musculus, Proc. Zool. Soc., 1865). They were described in B. rostrata by Drs Carte and Macalister (Trans. Roy. Soc., 1868), and by Mr J. B. Perrin (Proc. Zool. Soc., 1870). By the author, in this Journal, fully, in B. musculus (1871); in a toothed cetacean, Hyperoodon bidens (1871 and 1873); in the Greenland Right-Whale, Balæna mysticetus (1878); and in this whale, Megaptera longimana, in a preliminary note, at the meeting of the American Association for the Advancement of Science, at Philadelphia, in September 1884 (American Naturalist, February 1885). By Dr John Anderson, in Platanista gangetica (Anat. and Zool. Researches, 1878). By Sir William Turner, in Sowerby's Whale, Mesoplodon bidens (this Journal, 1885), with which he gives an account of his dissection of them in a fœtus of B. Sibbaldii, made in 1869-70. In the Narwhal, Monodon monoceros; the White Whale, Beluga; Globicephalus melas; and in the common Porpoise, Phocæna, I found these muscles to be present morphologically, but histologically represented by fibrous tissue, being functionally ligaments. But in Phocæna, the flexor carpi ulnaris was present in the fleshy condition. (This Journal, 1871, 1877, and at the Aberdeen Meeting of the British Association for the Advancement of Science, 1885.)

B. musculus. They were, on the internal aspect, three,—a flexor digitorum ulnaris, a flexor digitorum radialis, and the flexor carpi ulnaris; on the external aspect, one, an extensor communis digitorum. The proportions of the two flexors of the fingers were reversed, as compared with those of B. musculus, the ulnar flexor being about a third the size of the radial flexor, instead of larger than it, as in B. musculus.

The account of these muscles in Megaptera may be shortened by referring to the figures which I gave of them in B. musculus, here reproduced in Plate IV. figs. 13 and 14; and reference may be made to my detailed account of the muscles in B. musculus (this *Journal*, 1871). It was a mature B. musculus, 64 feet long, the pectoral fin 7 feet 8 inches long, the longest digit 33½ inches in length; while in this Megaptera the pectoral fin was 12 feet long, the longest digit 81 inches in length.

Flexor carpi ulnaris.—Belly does not spread like a fan as it does in B. musculus, but is thick and fusiform from the origin onwards. Origin entirely on the cartilaginous olecranon, abruptly from its distal edge, the aponeurosis reaching for about ½ inch forwards. Fleshy for 11 inches, being nearly half the length. Thickness of flesh at origin fully 1 inch, at middle ¾ inch. Belly covered by a thick aponeurosis of origin. Tendon of reception is continued on deep aspect of belly. Length of fleshy bundles about 1 inch, running obliquely between the two aponeuroses. The tendon, after 4 inches from the belly, is elliptical in section, lies edgeways to the ulna, is 1½ inch from the ulna, the total distance here from the ulna to the free upper border of the paddle, 6 inches. Tendon lies in a strong sheath, which also covers the belly, distinct from the aponeurosis of origin, and goes down to the upper border of the ulna. This is the strong fibrous curtain which I noted in B. mysticetus, absent in B. musculus. The blubber between the tendon and the free border of the paddle is much mixed with fibrous tissue, in longitudinal streaks. Distally the tendon gradually gets nearer the ulna, narrowing the fibrous curtain, expands in the last 3 inches, and is inserted entirely into the proximal border of the pisiform cartilage, not reaching quite up to the angle of the pisiform. The tendon at 4 inches from the flesh was nearly as bulky as the common tendon of the two flexors. The flexor carpi ulnaris is, therefore, as far as I could judge, as fully developed in Megaptera as in B. musculus, relatively to the flexors of the digits more so. It will serve to give increased resistance to this soft part of the paddle.

Flexor digitorum ulnaris.—This muscle resembles the same muscle in B. musculus, but is very much smaller. Belly, length 5 inches, flesh continued 1 inch farther on deep aspect of tendon ; greatest breadth, at middle, $1\frac{1}{2}$ inch ; thickness, $\frac{1}{3}$ to $\frac{1}{2}$ inch ; figure triangular. Origin from the ulna, doubtful if any fibres come so far forwards as from the humerus. Belly lies obliquely on the slope to the interosseous space. Tendon 14 inches before it joins tendon of radial flexor. Lies on interosseous slope of ulna and then for 5 or 6 inches sunk in the interosseous space, finally getting into same sheath as radial flexor. Tendon is about $\frac{1}{3}$ inch broad, and about twice the bulk of that of an average human plantaris muscle.

Flexor digitorum radialis.—Is about three times the size of the last muscle, but smaller than in B. musculus. Arises from the ulna as well as from the radius, beginning about 4 inches later than the ulnar flexor, length of belly 8 inches, the flesh running on the deep aspect of the tendon for 3 or 4 inches more. The tendon runs up in the belly like a septum, rendering it bipenniform, as in B. musculus. Bulk of belly about that of two thick fingers. Tendon, after 4 inches of pure tendon, in all 16 inches from the origin of the muscle, receives the tendon of the last muscle. This is at about the junction of the middle and distal thirds of the fore-arm. Tendon lies deep in the interosseous hollow, flattened sideways, about $\frac{1}{3}$ inch thick.

Dissection.—Each of these muscles lies in a sheath of fibrous tissue, $\frac{1}{4}$ to $\frac{1}{2}$ inch thick. The fleshy fibres partly arise from this sheath. The tendons are not very loose within the sheath, as if they did not move far. From their interosseous position and the great thickness of the sheaths these tendons, and fleshy parts too, might readily be overlooked. The first structure come upon, on slitting up the very thick aponeurosis of the fore-arm, on the flexor aspect, is a great nerve, as large as the forefinger, sur- rounded by loose areolar tissue within its sheath. Also in a sheath, along the radial border of the interosseous space, is a large artery, continued along the radial border of the common tendon.

Common Tendon.—The tendons of the two flexors unite in a common triangular expansion, the bulk of which appears to exceed that of the two tendons which form it. From this four tendons proceed, of about equal size, to the four digits. Breadth of the tendons along proximal half of the digits, 1 to 1½ inch ; on distal half, broader ; on the cartilages, 2 inches. They go to the ends of the digits. They have a fibrous covering, but not regularly formed thecæ and have no synovial covering, only areolar tissue between. Their function must be ligamentous, with traction exerted on them by the muscles.

The greater size of the radial flexor than of the ulnar flexor in Megaptera, in contrast with their proportions in B. musculus, is interesting as in adaptation to the greater size of digit II. in Megaptera. Although the two tendons unite in a common expansion, the radial flexor will exert its traction on the radial side.

Extensor communis digitorum.—Fleshy belly 18 inches in length, greatest breadth 1½ inch, being same length as in B. musculus, but only half the breadth. Tendon ⅝ inch in breadth ; in B. musculus was 1¼ inch. Is about size of human tendo Achillis. Forms large triangular expansion on distal half of carpus, which gives off four tendons at the proximal ends of the metacarpal bones. The division in B. musculus was earlier, at about the middle of the carpus. Tendon to digit III. the largest, that to digit V. the smallest. At middle of digit III. the tendon is twice the bulk of the entire tendon in the fore-arm ; breadth 2 inches, thickness ¼ inch. The tendons have attachments to all the bones and joints.

Function.—Here, then, we see in the great paddle of Megaptera the same muscles not half the size they have in the much smaller paddle of B. musculus, illustrating their rudimentary condition. The great increase of the bulk of the tendons on the digits is an illustration of their mainly ligamentous function. The traction exerted on them by the muscles will give some additional resistance on both aspects.

D

(B) The Hind Limb.

16. Table V., *showing, in inches, the Length, Breadth, and Thickness of the Pelvic Bone and the Femur in the Megaptera longimana.*

	Length.	Breadth.	Thickness.
	inches.	inches.	inches.
Pelvic bone (including cartilages) straight	9¼	...	...
" anterior portion (beak)[1] . .	5	...	...
" posterior portion (body) . .	6½	...	...
" ossified part	4¼	...	...
" anterior cartilage	2	...	...
" posterior cartilage	3¼	...	...
" at middle of beak, right . .	...	1⅛	⅜
" " " left . .	...	⅞	½
" at the promontory, right . .	...	2½	1
" " " left . .	...	2¼	1
" at middle of body, right . .	...	1½	⅝
" " " left . .	...	1¼	⅜
Femur (including ⅛-inch-thick perichondrium), right .	5	...	...
" " left . .	3¾	...	...
" greatest, near posterior end, right .	...	1¼	1⅛
" " " left .	...	1½	1
" at middle, right	...	1⅛	¾
" " left	...	1⅛	⅝
" at ½ inch from anterior end . .	...	⅝	½

17. The Pelvic Bone (figs. 15 and 16, *P.*).—As seen in the preceding Table, scarcely half of the length of the pelvic bone is ossified. This contrasts remarkably with the condition of the bone in Rudolphi's 44-feet-long Megaptera (English measure), also a male. In the full-sized figure which he gives (*op. cit.*, Taf. IV.) the length of the ossified portion is fully 9 inches, while neither of the cartilages is an inch in length. It contrasts also with the condition of the bone in my 50-feet-long B. musculus, in which the lengths of the corresponding parts are, the ossified part 8½ to 9 inches, the posterior cartilage ½ inch, the anterior cartilage about 1 inch. In form, the pelvic bone contrasts with that of this B. musculus in being less flattened and in having a much less extended promontory. The breadth at the promontory in the B. musculus is 4 to 4½ inches, in the Megaptera only 2¼ to 2½ inches. Nor is the promontory in Megaptera tipped with cartilage.

[1] The measurements of the anterior and posterior portions of the pelvic bone are taken from the middle of the outer border of the promontory to the tip of each.

The *acetabular cartilage* which I found in the Greenland Right-Whale (*loc. cit.*, figured in Plate XIV.), and which I find to be present in my 50-feet-long B. musculus, is entirely absent in this Megaptera. On raising the periosteum carefully at the promontory and from both surfaces near it, no cartilage of any kind is seen.

18. THE FEMUR[1] (figs. 15 and 16, *F.*).—The femur is entirely cartilaginous. On horizontal section in its whole length, the cartilage is seen to be traversed by the usual large Haversian canals, in that section divided transversely, at distances of $\frac{1}{8}$ to $\frac{1}{4}$ inch, less towards the tapering anterior end, wider apart towards the thicker posterior end. It is closely embraced by its perichondrial capsule, averaging $\frac{1}{8}$ inch in thickness, thinner behind thicker in front. The difference in length between the right ($3\frac{3}{4}$ inches) and the left ($5\frac{1}{4}$ inches, the cartilage proper 5 inches) is striking. The form is that of a pine-cone, a little flattened, so that the surfaces are inferior and superior, the borders internal and external. In Eschricht's figure of the fœtal cartilage, it has a somewhat pear-shape, with a pinched anterior third; in the adult he defines the form as "fast wie die einer menschlichen Kniescheibe," but his figure of it is longer and less pointed than a human patella. In this Megaptera (as seen in fig. 16, R. and L.) it presents two slight lateral projections on both borders with a constriction between. The projections are seen to correspond to the attachment of ligaments or other

---

[1] The presence of this bone in Megaptera was discovered by Eschricht. Writing in 1840 (*loc. cit.*, p. 136) he mentions having first found it in fœtal Humpbacks, male and female, as a cartilaginous nodule. In his figure (fig. 43) of the full size in a 78-inch-long fœtus, it is somewhat under $\frac{1}{2}$ inch in length, the pelvic bone $1\frac{3}{4}$ inch. He figures it (fig. 44, reduced to $\frac{1}{4}$th, the natural length would be nearly 2 inches) from a full-grown Humpback, the pelvis of which had been sent to him from Greenland, adding the important fact that, in this "erwachsenen Thiere," "Er war hier grösstentheils verknöchert." When the presence of a rudimentary femur in B. musculus was discovered by Professor Flower, C.B. (*Proc. Zool. Soc.*, 1865), it was in the condition of a cartilage, $1\frac{1}{4}$ inch long by $\frac{3}{4}$ inch broad, although the whale was 67-feet-long and a male. In the 64-feet-long B. musculus (*loc. cit.*, 1871, and plate vii. fig. 3) I found it mostly in an ossified condition, 2 inches in length, $1\frac{1}{4}$ in breadth, $\frac{5}{8}$ in thickness, ossified in the proximal $\frac{2}{3}$ of its length. But in my 50-feet-long B. musculus, also a male, the femur is entirely cartilaginous, $1\frac{1}{2}$ to $1\frac{3}{4}$ inch long by 1 inch broad. A preliminary account of the dissection of the hind limb of this Megaptera was read by me at the meeting of the British Association for the Advancement of Science, at Montreal, in August 1884.—(*American Naturalist*, February 1885).

fibrous bands. It may seem stretching comparison too far, but if the forms which I figured of the femur in the Greenland Right-Whale (*loc. cit.*, Plate XIV.) are looked at, it will be seen that these parts may be compared to the head, neck, and trochanter, the shaft represented only by the tapering anterior end.

The femur has no articular connection with the pelvic bone, joined to it only by ligaments posteriorly, $1\frac{1}{2}$ inch in length, allowing it to play loosely on the pelvic bone internal to and in front of the promontory. The anterior portion (beak) of the pelvic bone for a couple of inches in front of the promontory is concave on this aspect where the femur crosses it, the beak directed forwards and inwards, the femur forwards and outwards.

19. LIGAMENTS, OTHER FIBROUS STRUCTURES, AND MUSCLES CONNECTED WITH THE PELVIC BONE AND THE FEMUR.—Space will allow me to give here only a short account of the soft parts met with in the dissection. The muscles are very different from those which I figured in the male Greenland Right-Whale (*loc. cit.*, Plate XVI.), this in part resulting from the much greater development of the femur and the presence of a rudimentary tibia in the latter. The arrangement of the soft parts in Megaptera corresponds pretty closely to that in B. musculus, an account of which, with more complete illustrations, I hope to publish soon. Meanwhile I give the figures (figs. 15 and 16) showing the arrangement in Megaptera, and if these figures be referred to the following short notes of these parts may be understood. These figures are reduced to $\frac{1}{8}$th, from full-sized drawings which I made as the dissection proceeded. The chief point of interest was to ascertain to what extent function could explain the presence of so very rudimentary a structure as this representation of the femur in Megaptera.

Posterior Connections of the Pelvic Bone.—Passing across between the posterior ends of the pelvic bones is the *great interpelvic ligament* (*a.a.*, figs. 15 and 16). Attached for $1\frac{1}{2}$ inch to the bone, and about $\frac{1}{2}$ inch thick. It ties the pelvic bones together posteriorly, and supports the crura penis, which are involved in its tissue anteriorly, and entirely rest on it. Behind, it attaches the anterior part of the *levator ani muscle* (*b*, fig. 15), and more externally the inner part of the caudal

muscular mass (c.). Along the posterior edge of the great
ligament is seen the posterior edge of the *transversus perinei
muscle* (d.) mostly concealed by and attached to the ligament;
as broad and as thick as the palm of the hand and 6 to 8 inches
in length transversely. In the ring between this muscle and
the beginning of the levator ani muscle, is seen the *retractor
penis muscle* (e.e.), rope-like, right and left, passing forwards on
the under surface of the penis; composed of pale unstriped
muscular fibre, as in other cetaceans. On the great transverse
ligament and crus penis is seen the *ischio-cavernosus muscle*
(f.f.). This great muscle, 10 to 14 inches in length, 5 to 6
inches in breadth, and 3 inches in thickness at the middle,
extends still more on the dorsal aspect than on the under
aspect where it is seen in the figure; when split long-ways at
the middle, I estimated each half as equal in bulk to an average
human gluteus maximus. It has no direct connection with the
pelvic bone, its bundles passing entirely between the interpelvic
ligament, crura and corpora cavernosa penis. It is considerably
more developed, especially in breadth, in Megaptera than in B.
musculus. It and the transversus muscle are very much more
developed in Mysticetus, forming what I described and figured
(*loc. cit.*, figs. 13 and 14, *l.* and *m.*) as the great compressor
muscle and the posterior compressor muscle. The references
to the muscles of the hind limb in B. musculus are from my
dissection of the 50-feet-long one.

At the posterior part of the pelvic bone is the *posterior or
caudal muscular mass* (g.), as seen in section 2 inches behind
the bone, at least 6 inches broad by 3 inches thick. Backwards,
it begins to unite with its fellow in a median raphé after a course
of 18 inches, just behind the anus. Forwards, it is attached
internally to the outer part of the great interpelvic ligament
(c.), but mainly to the pelvic bone, to the end and for 3
inches on the outer side, on the outer border and both surfaces. Besides these fleshy attachments it sends forwards a
tendinous sheet worthy of particular notice. The outer part of
this sheet skirts the pelvic bone externally, and runs into the
anterior muscular (or tendinous) mass of the beak of the pelvic
bone; its middle and inner parts after covering and thus
strapping down the pelvic bone, pass, the middle part to be

attached to the outer edge of the femur, the inner part to form a fibrous sheet covering and adhering to the superficial surface of the femur. These latter parts, all however forming a continuous sheet, thus strap down the femur, and enable the caudal mass to act in part as a retractor of the femur, tightening it backwards and also outwards. This backward connection of the femur is strengthened by a deeper fibrous stratum, passing back from the femur to the pelvic bone and to the outer part of the great interpelvic ligament. Near the femur these posterior fibrous connections are ⅓ to ½ inch thick, farther back the deeper stratum is about ⅛ inch thick. The latter will serve purely as a ligament.

Anterior Connections of the Pelvic Bone.—The arrangement of the *anterior or trunk muscular mass* in Megaptera differs from that in B. musculus and still more from that in Mysticetus. In the latter, in a 33-feet-long *Mysticetus*, a great mass of flesh, 10 inches by 3 inches, came back to be attached, fleshy, to the beak, to the long nearly parallel femur, and to the tibia, separating into internal and external parts. In the *B. musculus* also there was a very large fleshy mass here, about 9 inches by 6, but most of it attached only to a great fibrous septum, to which also is attached a portion of the posterior caudal mass. To the anterior half of the beak was directly attached a mass of flesh 4 inches by 1 inch, and separately at the outer part, just in front of the promontory, a tendon, 1½ inch broad, which after a course of 1½ inch gave off a large lateral anterior muscle. In *Megaptera* this latter, 1¼ inch in breadth here, strengthened by a part arising on the outer side of the promontory, is the only structure I saw attached to the beak of the pelvic bone (fig. 15, *i.i.*). It was fibrous for 4 inches forwards, and was joined on its outer side by the part of the tendinous prolongation of the posterior caudal mass above noticed as skirting the pelvic bone externally. In connection with this difference in the soft parts attached to the beak in these two species of finners, is to be remarked the shortness of the beak, compared with the body of the pelvic bone, in Megaptera. Measured from the middle of the outer edge of the promontory, the lengths of the beak and the body are, respectively, in the 50-feet-long B. musculus 9 inches and 5½ inches; in the 64-feet-long B. musculus, $14\frac{1}{2}$ inches and 11 inches; while in this Megaptera the beak is 5 to $5\frac{1}{2}$, the body $6\frac{1}{4}$ inches.

The Superficial Interpelvic and Interfemoral **Aponeurosis.** —This great sheet of fibrous tissue is seen (fig. 15, *k.k.*) to pass across superficially between the pelvic bones and neighbouring parts, connecting them together, and supporting the parts of the penis. Breadth, antero-posteriorly, fully 12 inches, extending forwards in front of the pelvis to the transverse superficial muscle which exists there, and backwards to the beginning of the posterior third of the pelvic bone. Here it ends rapidly in a curved line with a median peak. If this edge is not natural the membrane is at least very thin from here back to the levator ani muscle, this space appearing after dissection as a perineal window through which the parts at the root of the penis are seen. This interpelvic aponeurosis may be regarded as a part of the general transverse aponeurosis of the region, specially thickened and attached where it lies between the pelvic bones and thigh bones. Thickness at the middle line, about $\frac{1}{4}$ inch, the part opposite the femur and fore part of the pelvic bone, $\frac{1}{2}$ inch. Its lateral connections at the pelvis are in three strata; the deepest attached to the pelvic bone; the middle, passing through between the pelvic bone and the femur, and blending with the deep tendinous tissue prolonged from the caudal mass; the superficial stratum, attached to the inner edge and superficial aspect of the femur, blending with its perichondrium. Behind the femur the aponeurosis joins the posterior fibrous connections of that bone; anteriorly it passes on the deep aspect of the anterior fibrous connections of the femur, to reach the inner edge of the anterior third of that bone and the deep longitudinal tendinous tissue, but it is not attached to the very apex of the femur or to the prolongation band, going across above these, and leaving them as more superficial parts. This great aponeurosis is composed of coarse transverse bundles of white fibrous tissue mixed with areolar tissue and fat. It was only after repeated examination with the microscope that I was satisfied it was nowhere muscular. In dissecting not very fresh cetacean tissues, streaks of brown-coloured blubber are sometimes extremely like muscle to the naked eye.

The Deep Ligaments and Retractor Muscle of the Femur. —The *posterior ligament* (figs. 15 and 16, *l.*), much the largest, resembles the letter Y reversed. Attached to posterior end of femur, undivided part 1½ inch long, as broad as forefinger but not so thick. External division flattened, ¾ to 1 inch broad, directed backwards and a little outwards, 1 to 1½ inch long, attached to pelvic bone about 2¼ inches from femur. Internal division conical form, slopes backwards and inwards for 3 to 4 inches, attached to great interpelvic ligament and crus penis by a 2-inch-broad base, beginning 1 to 1½ inch internal to the pelvic bone.

On cutting into this limb of the posterior ligament, it is seen to be hollow and to contain a muscle, composed of red striped fibre. This *retractor femoris muscle* (fig. 16, right side, *r.m.*) is from 2¼ to 3 inches in length; in breadth, at the base 1½ to 2 inches, at the middle 1 inch; thickness at the base about 1 inch. The enclosing sheath (the ligament) is $\frac{1}{10}$ inch thick. The chief origin of the muscle is, for its outer half, from the interpelvic ligament, and for its inner half from the ¼ to ½-inch-thick fibrous wall of the crus penis. There it comes in close relation with fibres of the ischio-cavernosus muscle, but the two muscles diverge immediately. Part of the bundles arise from the inner surface of the sheath at its base, and are inserted into the sheath farther forwards, especially towards the apex, the central part ending in a short tendon which soon becomes identified with the central part of this limb of the ligament. The above applies to the ligament and muscle of the right side, that on which the femur is largest. On the left side, the internal limb of the ligament and the contained muscle are much less developed. This muscle will pull the femur backwards and a little inwards, while the enclosing sheath will serve as a ligament checking forward movement. Considered in relation to the size of the bone on which it acts, this is a large muscle, having a bulk of flesh say equal to the thumb modelled into a cone. The posterior ligament is the great one in all these cetacea possessing a femur, but I have not found in B. musculus or in Mysticetus anything corresponding to the inner limb of the ligament and its contained muscle.

External Ligament of the Femur (fig. 15, *m.*).—A flattened triangular band, 1½ inch long, attachments towards promontory and outer side of head of femur; at middle $\frac{3}{4}$ inch broad, $\frac{1}{10}$ inch thick, but about half that size on left side.   Its pelvic attachment is at what would be the acetabular cartilage in Mysticetus and in B. musculus.   There is an elevation of the bone here in Megaptera, but rather to the inside of where the acetabular cartilage is in these other whales.   The ligament is attached at the outer side of this elevation on the left side, at its inner on the right.   On the left side the ligament, directed inward and forward, will check gliding movement of the femur in these directions.   On the right side, owing also to the more outward position of the femur, the ligament turns round the outer edge and goes on to the outer part of the superficial surface of the femur, and is so placed that it checks gliding movement in the outward direction, and rotation inwards.   On the left side an intermediate ligament, stronger than the external ligament, is seen, associated more with the posterior ligament than with the external.   Not present on right side.

The Anterior Fibrous Connections of the Femur.—The great anterior band of the femur is attached not only to the apex, but by thinner continuations to each side of the anterior third of the femur, as far as the anterior lateral tuberosity (fig. 15, *n.*, right side).   In the figure, on the left side, these continuations are removed, bringing into view (fig. 15, *o.*) a band which arose from the deep surface of the femur, opposite to that tuberosity, $\frac{3}{4}$ inch broad, flattened but a thick strong band, stronger on the right side than on the left.   It is a deeper stratum of the fibrous tissue at the outer part of the great anterior band, separated, seemingly, by its being connected externally with the fibrous prolongation from the caudal mass, a portion of which is seen joining it in the figure.

The great anterior band may be termed the *femoral prolongation band*, regarding it as representing a continuation of the femur, like the tibial band which appears to represent a continuation of the tibia in Mysticetus (*loc. cit.*, fig. 18, *k.*). Arising at and near the apex of the femur, it has a size of $1\frac{3}{4}$ inch broad by $\frac{1}{4}$ to $\frac{1}{3}$ inch thick, oval in section, thicker internally than externally; passes forwards for about 15 inches, expanding, and ends by joining the fibrous tissue at the posterior part of the large superficial transverse muscle, which there supports the

E

prepuce on the anterior half of the penis. Although not noted
during the dissection, the edges, I think, were joined by, or gave
off, a fibrous expansion, but the band stood out prominently as
a long flat tendon-like structure, and, as above noted, lay superficial to the great transverse interpelvic aponeurosis.

Summary of the Connections of the Femur.—(*a*) The *purely
fibrous* connections are, *posteriorly*, the deep posterior ligament
to the pelvic bone; and superficial to it, the fibrous stratum to
the pelvic bone and interpelvic ligament. These will offer
strong resistance to over-advancement of the femur. *Anteriorly*,
the prolongation band. The anterior attachment of this band
not being to bone, the resistance offered by it to retraction will
not be very definite. *Internally*, to its fellow, by the great
transverse aponeurosis. This will offer strong resistance to
outward movement. *Externally*, the external lateral ligament
to the pelvic bone; and the adhesion of the longitudinal
tendinous tissue on the outer side of the pelvic bone. (*b*)
Muscular influences.—From the caudal muscular mass, by those
parts of its anterior tendinous prolongation which are attached
to the hinder end and outer border of the femur. Will tend
to pull the femur backwards and a little outwards when the
pelvic bone is being retracted. The special retractor muscle,
which will pull the femur backwards and a little inwards, when
a tight condition of the interpelvic ligament and crus penis gives
the muscle a fixed point to pull from. The only muscular
action on the femur, therefore, appears to be retraction, and
the chief ligamentous resistance is against advancement.

Exact Position of the Femur.—On the right side, that of the
larger femur, about the posterior half of the femur lies on the
pelvic bone. On the left side, the like decussation of the axes
of the beak and of the femur, leaves but the apex of the femur
in front of the beak. The right femur lies in the general hollow
of the beak, with their periosteum and loose tissue between.
On the left side, in addition, the middle stratum of the transverse
aponeurosis was noted as lying between, the left femur being
placed somewhat more internally than the right, as represented
in the figure. Now, in the ligamentous preparation, the parts
being quite loose, it looks as if the left femur had lain in the
same position as the right, allowing for its being shorter. The

laxity of the ligamentous connections of the femur in Megaptera, compared with those of Mysticetus and B. musculus, is striking.

20. FUNCTION.—The more the connections of the femur in Megaptera are examined the less easy does it seem to give a functional explanation of its presence. It might be looked on as serving a sesamoid function, but it does not play on cartilage, and does not give the mechanical advantages of a sesamoid. It has even less muscular connection than the small oval femur in my 50-feet-long B. musculus. Mysticetus has also a rudimentary tibia, Megaptera a femur only, B. musculus a still more rudimentary femur, and B. borealis, as I find, none at all.

21. EXPLANATION OF PLATES III., IV., AND V.

Fig. 6. Left paddle of Megaptera, flexor aspect, reduced to $\frac{1}{21}$. The epiphyses of the humerus and fore-arm are seen. The epiphyses of the fore-arm and metacarpal bones, at the carpus, are to be distinguished from the carpal bones proper. The causes of the nine nodes on the radial border, shown in Part I., Plate I., are seen; and of the minor undulations on the ulnar border, near the tip. The fitting of the alternating nodes and hollows of the digits, the position of the joint in the nodes, and of the joints in the terminal cartilages, are represented.

Fig. 7. Left scapula of Megaptera, turned round to show the dorsal surface, with its very low spine, s. f, prescapular fossa; a, anterior angle; p, posterior angle; c, rudimentary coracoid; e, low elevation; reduced to $\frac{1}{21}$.

Fig. 8. View of glenoid cavity of same scapula. c, rudimentary coracoid, not yet completely united to the scapula.

Fig. 9. Dorsal view of section of left carpus of Megaptera with portions of radius and ulna and their epiphyses, and portions of the metacarpal bones and the epiphysis of each; reduced to $\frac{1}{12}$. r, radiale; i, intermedium; u, ulnare; p, pisiform; 3, os magnum; 4, unciform bone. Ossification is seen in the radiale and ulnare, and in the epiphysis of the radius and of the ulna. These ossifications are seen only on section. The dotted line is where the pisiform and the epiphysis of the ulna are not completely separate. The proximal synovial cavity at the os magnum, 3, is represented.

Fig. 10. Terminal cartilage, with distal phalanx, of digit III.; a, the joint in the cartilage; reduced to $\frac{1}{5}$.

Fig. 11. Terminal cartilage, with part of distal phalanx of digit II. (Index digit). a, its first joint; b, its second joint. Reduced to $\frac{1}{5}$.

Fig. 12. Longitudinal vertical section of the terminal cartilage of digit II., showing the two joints a and b; the proximal reaching both surfaces; the distal seen only on section; reduced to $\frac{1}{5}$. The flexor aspect of fig. 12 is that next to fig. 11. The radial border of figs. 10 and 11 is towards fig. 12.

Figs. 13 and 14. Pectoral fin of the 64-feet-long B. musculus, reproduced from this *Journal*, 1871, for comparison with that of Megaptera; and for the muscles, flexor and extensor. Reduced to $\frac{1}{16}$.

Fig. 15. The pelvic bone and femur and **their** muscular and fibrous connections in this male Megaptera longimana; reduced to $\frac{1}{6}$. The dissection carried deeper on the left side. *P*, pelvic bone; *F*, femur. The dotted line shows the position of the right pelvic bone; *a*, great interpelvic ligament, at its attachment to pelvic bone, *b*, part of levator ani muscle; *c*, part of the **caudal** muscular mass; *d*, transversus perinei muscle, only the posterior border seen; *e.e.*, retractor penis muscle, right and left; *f*, ischio-cavernosus muscle, lying on interpelvic ligament and crus penis. The dotted lines show its position covered by the interpelvic aponeurosis; the **inner** dotted line shows its line of termination on the under aspect of the corpus cavernosum penis; *g*, attachments of caudal muscular **mass to pelvic bone** and interpelvic ligament; *h*, prolongations of its tendon to the femur and along outer side of pelvic bone; *i.i.*, attachments of anterior muscular mass to pelvic bone, in Megaptera only by tendon; *k.k.*, great interpelvic and interfemoral aponeurosis; *l*, posterior ligament of femur, bifurcated. A portion of the internal limb is seen on the right side; *m*, external ligament of femur; *n*, femoral prolongation band On left side the thinner lateral parts of that band removed, showing *o*, separate part attached on deep aspect of femur and joined by part of tendinous prolongation from caudal mass.

Fig. 16. Pelvic bone and femur and their ligaments reduced to $\frac{1}{6}$. The right femur is seen to be larger than the left. The letters refer to the same parts as in fig. 15; *r.m.*, on right side, retractor femoris muscle, contained in the inner limb of the posterior ligament.

PART III.

THE VERTEBRAL COLUMN.

VERTEBRAL COLUMN.[1]

THE following Table gives the number of vertebræ in each region, with the length of each region and other measurements, in Megaptera, and, for comparison with it, in B. musculus and in B. borealis:—

TABLE I.

	Carcass.	Vertebral Column.		Cervical Region.			Dorsal Region.			Lumbar Region.			Caudal Region.			Leaving for Head.	
	Length in Feet.	No. of Vertebræ.	Feet.	Inches.	No. of Vertebræ.	Feet.	Inches.	No. of Vertebræ.	Feet.	Inches.	No. of Vertebræ.	Feet.	Inches.	No. of Vertebræ.	Feet.	Inches.	Feet.	Inches.
Megaptera	40	52	28	9½	7	1	7	14	7	½	10	7	2	21	13	...	11	2¼
B. musculus,	50	62	37	...	7	1	7	15	8	7	15	11	10	25	15	...	13	..
B. borealis, .	36	56	28	...	7	1	5	14	6	2	14	9	6	21	10	11	8	...

[1] In making the following observations on the vertebral column of Megaptera, those of B. musculus and B. borealis were placed beside it for comparison. In this way an appreciation of its distinctive characters could be made. With this view a comparison is given with the corresponding parts in B. musculus, in so

F

Number of the Vertebræ.—Megaptera has the same number
of dorsal vertebræ as B. borealis and one less than B.
musculus, but the striking difference is in the lumbar region,
Megaptera having only 10, while B. borealis has **14**, and B.
musculus **15**. This gives great shortness to the abdominal
region in Megaptera. The caudal region is long compared
with that region in B. musculus and in B. borealis in proportion
to the total length. The 13 feet of caudal region in Megaptera
against 11 feet in B. borealis, while both have 21 caudal
vertebræ, is not owing to greater length of the bones in
Megaptera but to the great length of the intervertebral fibro-
cartilages in Megaptera. The contrast between Megaptera and
the other two finners in this respect, seen in dissection, was
apparent also in the dorsal and lumbar regions, but was striking
in the caudal region. Megaptera is thus characterised by a
short " trunk " and a long, robust, and flexible tail, surmounted
by a large tail fin. These proportions of the vertebral column
are taken from the three articulated columns placed beside each
other, the proportions carefully assigned in articulation.

Number of Caudal Vertebræ.—The precise number of the
caudal vertebræ must often remain uncertain. In this 50-feet-
long B. musculus, what would have come out of the maceration
trough as the last vertebra is about 1 inch broad and $\frac{1}{2}$ inch
long. But behind it is a conical cartilage, $\frac{1}{2}$ inch long in the
now dried condition, and on the upper aspect of this, at its
middle, is a rounded bony nucleus only $\frac{1}{8}$ inch in diameter.
What is present as the last caudal vertebra in this Megaptera
is a piece of bone about the size of a common nut compressed
into a somewhat cubical form ($\frac{3}{4}$ inch broad, $\frac{1}{2}$ inch in height
and length, but part of the length is evidently broken off).
There may have been a cartilage behind it as in this B.
musculus, and still more so in this B. borealis, the last vertebra
of which present is $1\frac{1}{2}$ inch broad, $\frac{7}{8}$ inch long, and $\frac{6}{8}$ inch high.
There can be absolute certainty only when the very end has

far as comparison is of interest. As this is sufficient to bring out the characters
of Megaptera, only occasional references are made to B. borealis. The comparison which I at the same time made between B. musculus and B. borealis was
interesting, as bringing out numerous differences between these two species, but
this subject I must reserve for some future communication. The parts relating
to B. musculus are placed within brackets [ ] for facility of reference.

been carefully dissected, but the circumstances attending the dissection of large whales are generally not very favourable.

Explanation of some of the Measurements in **Tables II.** *and III.* (pp. 64-67.) —
No. 1. Vertically from top of spinous process to back part of the body, or, when spine very sloping, to level of under surface of body. In neck to level of lowest part of body.

No. 3. By callipers, from within spinal canal, along middle line of process. From within canal is the only definite point below. In the cervical vertebræ the thickness of the lamina is then deducted, as this had been done in the measurements given in my table of the cervical vertebræ in this *Journal* in 1872, pp. 20, 21.

No. 9. When, as in B. musculus, the measurement would differ according as taken from where the process leaves the fore part or the back part of the body, it is taken from a line drawn between these two points.

No. 13. With callipers, towards the middle of the ends. At the margins it might have been ¼ to ½ inch less.

No. 14. At anterior end, and on the edges of the epiphyses. At the middle would be less reliable, as some have a ridge, some a hollow there. At the epiphysis keeps clear of the chevron tubercles.

No. 15. At anterior end, and on the epiphyses. Just behind the epiphyses might give ¼ to ½ inch more, but the epiphysis is the least variable part, and will form the true end of the consolidated bone.

No. 17. Taken, like No. 16, at the middle. Taken below or above the transverse process according as it is before or behind the transition vertebra ; the 7th of Megaptera, the 6th of B. musculus.

No. 18. This measurement may be influenced to the extent of about ⅛ inch by a superior median ridge, or by the top of the arch being a little more or less pointed.

No. 20. Between upper edges at fore part, being the most prominent part anteriorly. In the neck, where the upper margin becomes the outer, the measurement is between the outer margins of the anterior processes.

No. 21. This measurement, taken with No. 1, shows the contribution made by the chevron bones to the two-edged knife, or "razorback," form of the caudal region in B. musculus, compared with Megaptera.

BODIES OF THE DORSAL AND LUMBAR VERTEBRÆ.

4. LENGTH, BREADTH, AND HEIGHT OF THE BODIES.[1]—On referring to Table II. it will be seen that the largest bodies are those of the first three or four caudal vertebræ. This part is the foundation of the great propelling organ. It is at the junction of about the posterior with the middle thirds of the entire carcase in Megaptera. The greatest breadth (11 inches) is attained by the 2nd and 3rd; the 1st and 4th only ⅛ inch

[1] The epiphyses of the bodies are still ununited throughout the spine. The ends of the spinous and transverse processes, and the upper margins of the articular processes, all show evidence of unfinished ossification.

TABLE II.—*Measurements of the Vertebræ of Megaptera longimana, given in inches.*

	Cervical, 7.							Dorsal, 14.													
	Atlas.	Axis.	3	4	5	6	7	1	2	3	4	5	6	7	8	9	10	11	12	13	14
1. Extreme height,																					
2. ,, width,																					
3. Spinous process, length,																					
4. ,, ,, breadth at middle,																					
5. ,, ,, breadth at end,																					
6. ,, ,, thickness at middle,																					
7. ,, ,, thickness at middle of end,																					
8. ,, ,, greatest thickness at end,																					
9. Transverse process, length,																					
10. ,, ,, greatest breadth,																					
11. ,, ,, least breadth (neck),																					
12. ,, ,, thickness at middle,																					
13. Body, length,																					
14. ,, height,																					
15. ,, width,																					
16. Pedicle, breadth at narrowest part,																					
17. ,, thickness at middle,																					
18. Spinal canal, height,																					
19. ,, greatest width,																					
20. Between anterior articular processes, at front,																					
21. Extreme height with chevron bones,																					

TABLE II.—*Measurements of the Vertebræ of Megaptera longimana, given in inches*—continued.

	Lumbar, 10.										Caudal, 21.																					
	1	2	3	4	5	6	7	8	9	10	1	2	3	4	5	6	7	8	9	10	11	12	13	14	15	16	17	18	19	20	21
1. Extreme height,																															
2. " width,																															
3. Spinous process, length,																															
4. " " breadth at middle,																															
5. " " breadth at end,																															
6. " " thickness at middle,																															
7. " " thickness at middle of end,																															
8. " " greatest thickness at end,																															
9. Transverse process, length,																															
10. " " greatest breadth,																															
11. " " least breadth (neck),																															
12. " " thickness at middle,																															
13. Body, length,																															
14. " height,																															
15. " width,																															
16. Pedicle, breadth at narrowest part,																															
17. " thickness at middle,																															
18. Spinal canal, height,																															
19. " greatest width,																															
20. Between anterior articular processes at front,																															
21. Extreme height with chevron bones,																															

TABLE III.—*Measurements of the Vertebræ of Balænoptera musculus, given in inches.*

	Cervical, 7.							Dorsal, 15.							
	Atlas.	Axis.	3	4	5	6	7	1	3	5	7	9	11	13	15
1. Extreme height,	12½	13¼	10¾	11	11¼	12¼	13¼	14	16½	19½	21½	22½	22¾	23½	24
2. Extreme width,	24¾	29¾	27¼	26½	26¾	26	26¼	26¾	25½	29	32¾	33½	34¾	35½	35¼
3. Spinous process, length,	¼	½	¾	⅞	½	1¼	1⅞	3¼	5½	9	9¾	11¼	11¾	13	13½
4. " " breadth at middle,	.	.	.	.	.	.	.	1½	3	4¾	5¼	6	6	5½	5¼
5. " " breadth at end,	.	.	.	.	.	.	.	¾	3¾	5½	6¼	7⅜	7⅝	7⅞	8¼
6. " " thickness at middle,	.	.	.	.	.	.	.	½	⅜	½	¾	1¼	¾	⅞	¾
7. " " thickness at middle of end,	.	.	.	.	.	.	.	⅜	½	½	⅝	⅞	⅞	⅞	⅞
8. " " greatest thickness at end,	.	.	.	.	.	.	.	⅜	¼	½	⅞	⅞	⅜	1	1¼
9. Transverse process, length,	6½	10¾	9½	9	9	8½	8½	9	8½	10	11½	11¾	12	12¼	12¾
10. " " greatest breadth,	.	.	.	.	.	.	.	1¼	4	5¾	6¾	7½	7½	7	6½
11. " " least breadth (neck),	.	.	.	.	.	.	.	1¼	2½	4	3¾	4½	4½	4¼	4¼
12. " " thickness at middle,	1½	1¼	½	½	½	½	1	2	1⅞	2⅝	1⅜	1⅝	⅞	⅞	⅞
13. Body, length,	3	2¾	1¼	1¾	2	2¼	2¾	3¼	4¾	6	7	7½	7¾	7¾	8¼
14. " height,	2½	6¾	7	7⅞	7⅞	7⅞	7⅞	7½	7¾	7	7½	7½	7¾	7¾	8
15. " width,	.	.	11⅞	11½	11¼	11½	11¼	11½	11	10	9⅞	10¼	10⅜	10¾	10⅝
16. Pedicle, breadth at narrowest part,	.	.	2½	2½	2⅞	3¼	3½	3¾	4	4¼	4½	4¼	4¼	4¼	4½
17. " thickness at middle,	.	.	⅝	⅞	⅝	¾	¾	⅝	1¼	2¾	¾	½	⅞	½	½
18. Spinal canal, height,	8	4¾	3¾	2¾	2⅞	2⅞	2⅞	3	3½	4⅜	4⅞	4½	4⅜	4	3⅞
19. " " greatest width,	4	6¾	6⅜	6⅞	7	7	7	7 1/16	6⅜	5¼	4⅜	3¾	3½	3¾	3¼
20. Between anterior articular processes, at front,	.	.	9½	9¼	9	9½	9½	10	9½	8	4½	3⅜	3½	3	2¼
21. Extreme height with chevron bones,	.	.	.	.	.	.	.	.	.	.	.	.	.	.	.

TABLE III.—*Measurements of the Vertebræ of Balænoptera musculus, given in inches*—continued.

	Lumbar, 15.							Caudal, 25.												
	2	4	6	8	10	12	14	1	3	5	7	9	11	13	15	17	19	21	23	25
1. Extreme height,																				
2. Extreme width,																				
3. Spinous process, length,																				
4. „ breadth at middle,																				
5. „ breadth at end,																				
6. „ thickness at middle,																				
7. „ thickness at middle of end,																				
8. „ greatest thickness at end,																				
9. Transverse process, length,																				
10. „ greatest breadth,																				
11. „ least breadth (neck),																				
12. „ thickness of middle,																				
13. Body, length,																				
14. „ height,																				
15. „ width,																				
16. Pedicle, breadth at narrowest part,																				
17. „ thickness at middle,																				
18. Spinal canal, height,																				
19. „ greatest width,																				
20. Between anterior articular processes, at front,																				
21. Extreme height with chevron bones,																				

less; the greatest height (10¼ inches) by the 2nd, 3rd, 4th, and 5th; the greatest length (8⅝ inches) by the last lumbar and two first caudal, the 3rd and 4th only ⅛ inch less. An eighth of an inch in such measurements may be only incidental. Thus the 2nd caudal has the greatest size in all directions—breadth 11 inches, height 10¼, length 8⅝; and the 3rd and 4th are scarcely less; the 1st only a little less.

From this region the bodies diminish forwards and backwards. In the diminution backwards the height and breadth keep about the same proportion in the first half of the caudal region, and in the last half are more nearly equal. The length remains less than the height or breadth until the fourth last caudal is reached. In the diminution forwards, the decrease in *breadth* is not great—about 1 inch along the lumbar region (10⅝ to 9⅞); along the dorsal region, 1 inch at the 3rd dorsal (9 inches), and at the 1st dorsal ⅝-inch more (to 8⅜-inches), a total diminution of about 2½ inches from the greatest breadth. In *height* the diminution goes on gradually, from the 10¼ inches on the 2nd caudal, to 7 inches at the middle of the dorsal region, and remains at this on to the 1st dorsal, a total diminution of 3¼ inches. In *length* the diminution is great—from the 8⅝ inches of the 10th lumbar, it has fallen to 6⅝ on the last dorsal, and to only 2¾ on the first dorsal. On the posterior half of the dorsal region it is only 1 inch. On the anterior half it is more rapid, from 5⅜ inches to 2¾, being fully one-half. The shortening of the bodies backwards from the great vertebræ is much more rapid than forwards, the length of the 17th and 18th caudal being about equal to that of the 1st dorsal, a distance backwards of 10½ feet (including 15 to 16 vertebræ, from the 2nd caudal), and forwards a distance of 13½ feet (including 23 vertebræ, from the 10th lumbar to the 1st dorsal).

The shortness of the bodies is seen when they lie alongside those of the other finners. The following are the measurements in inches, of the 13th dorsal in the three:—

	Length.	Height.	Breadth.
Megaptera, . . .	6	7½	9½
B. musculus, . . .	7	7¾	10¾
B. borealis, . . .	6	5⅛	7⅛

In Table II. it is seen that, all along the column, the length is considerably less than the height, back to the 17th caudal, where they become about equal.

[In *B. musculus* the largest bodies are those from the 3rd to the 7th caudal, the 5th on the whole the largest. Its height is 10¾ inches, length 10 inches, but the 7th has the greatest breadth, 12½.

The greatest vertebral body in Megaptera is the 2nd of its 21 caudal vertebræ, the 33rd of its 52 vertebræ; and it has in front of it nearly 17 feet of the vertebral column, behind it, 11½ feet. In B. musculus the greatest body is that of the 5th of its 25 caudal vertebræ, the 42nd of its 62 vertebræ; and it has in front of it 26 feet of the column, behind it, 10½ feet. In B. musculus the length is as great as the height in the posterior dorsal region (9th to 13th); along the lumbar region, less than the height only by ¼ to ½ inch; in the caudal region the difference may be only slightly greater back to the 11th, where the length becomes less than the height by 1¼ inch, and more backwards, except on the four or five posterior vertebræ where the difference between the height and the length is not great. In Megaptera the length is less than the height on the above-mentioned dorsal vertebræ (9th to 13th) by from 1⅛ to 1½ inch; along the lumbar region by an average of 1½-inch, in contrast with the ¼ to ½ inch in B. musculus. In *B. borealis* the length is even greater than the height, from the 9th dorsal back to the 7th caudal.]

5. THE EPIPHYSES.—The lengths of body given include the epiphyses. On the 2nd or 3rd caudal the epiphyses, at midway between the transverse process and the pedicle, are ½ inch thick. [On the corresponding and neighbouring vertebræ of the B. musculus, the epiphyses are 1 inch thick; in the B. borealis they are ½ inch thick.] In front of the last two lumbar, the thickness has diminished to ⅜ inch; at the first dorsal, to about ⅔. Backwards, along the caudal region, the epiphyses remain undiminished in thickness to the 10th, after which the bevelling of the edges, especially of the posterior epiphysis, renders it not easy to measure the thickness, but back even to the 17th caudal, the anterior epiphysis is still ½ inch thick. After the 10th caudal, the posterior epiphysis is somewhat thinner than the anterior. After the 10th, the back of the bodies becomes convex, the front remaining more nearly flat. This is due to the bevelling off of the hinder edge of the posterior epiphysis in the rapid tapering of the hinder vertebræ of the tail. These measurements are taken at the unfinished abrupt edges of the epiphyses, but 1 inch to 1½ further in,

where the annular platform is most raised, the thickness of the
epiphyses is greater. This is well seen when the epiphysis
happens to be broken. The posterior epiphysis of the 2nd
lumbar, $\frac{3}{8}$ inch at the edge, is $\frac{6}{8}$ inch at $1\frac{1}{4}$ inch in from the
edge. The anterior epiphysis of the 11th caudal, scarcely $\frac{1}{2}$ inch
at the abrupt edge, is almost 1 inch further in. On the 15th,
16th, and 17th caudal, which admit here of being measured
throughout, the anterior epiphysis is $\frac{1}{8}$ inch thicker than the
posterior of the same vertebra, and is not thicker internally than
at the edge. From the 12th to the 17th caudal, the bodies can
be exposed and are seen to be convex on both aspects where
covered by the epiphyses, rendering the deep surface of both
epiphyses cupped. The length of the bodies was measured from
the annular elevation, giving the greatest length.

6. FORM OF THE BODIES—*Markings on the Ends.*—Megaptera
differs from B. musculus in the appearances presented by the
ends of the bodies. In Megaptera the streaked annular platform for the capsular part of the intervertebral disc is broader
in proportion to the enclosed pulp-area, is more raised, and has
its lines more pronounced than in B. musculus. There is a
groove between them in Megaptera owing to the elevation of
the capsular ring especially at its inner part, while in B. musculus the central area stands out more abruptly owing to the
flatness of the ring. The greater relative breadth of the ring
in Megaptera is seen throughout the spine, and increases backwards. At the 6th lumbar vertebra, the proportions of the
ring to the pulp-area are, in Megaptera, $1\frac{3}{4}$ and $6\frac{1}{2}$ inches; in
B. musculus, $1\frac{1}{2}$ and $7\frac{1}{2}$ inches; at the 6th caudal, in Megaptera,
$2\frac{1}{2}$ and $5\frac{1}{2}$ inches; in B. musculus, $1\frac{3}{4}$ and $7\frac{3}{4}$ inches. [In B.
musculus the ring is flat, becoming a little convex in the caudal
region. B. borealis in these respects resembles B. musculus in
contrast with Megaptera.]

Form of the Ends.—The ends of the bodies in front of the
caudal region are flat. They may undulate a little from the
slight projections at the pedicles and transverse processes, and
appear slightly convex from the falling away external to the
elevated part of the capsular platform, but on the whole the
end is flat. After the 1st caudal, one on both ends become
convex. From the 2nd to the 8th the convexity is not great

and is about equal on both ends. The 9th and still more the
10th, have the posterior end more convex than the anterior; the
11th, the first post-chevron vertebra, very much so. The 12th
and 13th, more especially the 13th, become decidedly bi-convex,
the posterior surface still the most convex, their curvatures
much resembling those of the human crystalline lens. The concavity is not due merely to bevelling but is seen over the central
pulp-area also. Behind these two the plano-convex form is
resumed, the anterior end nearly flat, the convexity of the
posterior end less than that of the biconvex vertebræ. The two
biconvex bodies are those which follow, with one transition
vertebra between, after a complete neural arch and chevron
bones have ceased. This form of the bodies must give great
freedom of motion to this part.

The caudal vertebræ, from and after the 14th, instead of having the sides convex both ways, like the three just in front, are
pinched at the middle, rendering the sides concave vertically at
the middle. This gives a four-cornered shape, in antero-posterior view, to these vertebræ, in striking contrast to the cir-
cular form of the vertebræ in front of them.

[In *B. musculus* the vertebræ continue flat-ended back to the
15th caudal (the first after the disappearance of a complete neural
arch), the posterior end of which is somewhat convex. This increases on the 16th, and the plano-convex form is continued on the
vertebræ behind, but there are no bi-convex vertebræ in this B.
musculus. The four-cornered form begins on the 18th caudal, but
is not very well marked till the 20th. The pinching at the sides is
rather in the form of an obliquely upward and backward fossa than
that of the vertical concavity of the middle of the side presented by
Megaptera. In *B. borealis* the 12th and 13th caudal, and to a less
extent the 14th, are bi-convex.]

Superior and Inferior Median Ridges of the Bodies.—
The sharp *superior median ridge* seen on the bodies of the
cervical vertebræ flattens down on the anterior dorsal into a
low broad median convexity. Increasing as we go back, this
elevation at the 5th dorsal has assumed a definite form 2
inches in breadth with a definite oval foramen, it may be
more than one foramen, on each side of it. At the
beginning of the lumbar region it is more raised and narrowed to 1½ inch, and bounded by a more elongated entrance
to the foramen. From the 7th to the 10th lumbar the

ridge rises abruptly, like a finger laid along the middle of
the body, the foramina in some (7th and 10th) communicating
freely below it. This longitudinal bridge on the 8th, now
broken, has been constricted, or even wanting at the middle.
On the first five caudal it stands up as a triangular ridge, with
the depression and foramen, or foramina, at each side, narrowing backwards to the 5th, where it is a sharp well-marked ridge.
On the 4th caudal it has again been undermined, so as to form
a bridge, so narrow that it has given way. On the 6th caudal,
and on the four succeeding vertebræ, those with neural arch
complete, it has entirely disappeared, the floor of the canal
being smoothly concave from side to side. After an interval of
four vertebræ (11th to 14th), which have a large median foramen, the median ridge reappears, on the 15th and succeeding
vertebræ, with a foramen or two on each side of it.

This ridge must belong, as I found in the neck of *B. musculus*,
to the superior common ligament of the bodies. The 6th
caudal, on which it ceases as a ridge within the canal, is the
last vertebra with a pronounced transverse process.

[This pronounced superior median ridge is a character of Megaptera,
as compared with B. musculus and B. borealis. In *B. musculus*, it is
seen in the neck, and very faintly on the two or three anterior dorsal,
not at all on the other dorsal, and very faintly on the lumbar and
anterior caudal; in the posterior part of the lumbar region as a
broad low convexity, but not at all projecting as in Megaptera. In
B. borealis this ridge is even more completely absent, faintly per-
ceptible to the finger in the posterior lumbar region only.]

Inferior Median Ridge.—This sub-vertebral median ridge,
seen especially in the *lumbar region* of Cetacea, is strongly
marked both in depth and breadth about the middle of the
lumbar region in Megaptera. Beginning on the 1st lumbar it
is there broad and long; on the 2nd it is sharp, but still has
the deeply concave outline. From the 3rd to the 8th this
crest projects so much as to give the body, on side view, a
convex outline below, strongly marked on the 5th and 6th. On
the 9th and 10th it is well marked, but the outline is again a
little concave. Viewed from below, the crest is seen on the
4th lumbar to broaden out posteriorly; on the 5th, its posterior
⅔ form a wide triangular surface, 2 inches broad behind; on
the 6th, the whole ridge is broad, ⅔ inch at the middle,

expanding at both ends. On the next four again it is a sharp ridge, broadening out on the posterior $\frac{1}{4}$ or $\frac{1}{3}$ into a triangular surface, more marked as we go back, but on the 10th it shows no tendency to the bifurcation presented by the 1st caudal for its chevron bone.

The subvertebral ridge shows itself on the *dorsal region* from the 5th to the 12th. The first five dorsal bodies are, on side view, rather convex below, the 4th decidedly so, but, unlike the cervical, have no distinct median ridge. Behind the 5th they are a little concave below, notwithstanding the crest, except the 7th and 9th, which are nearly level below. Seen from below, this ridge on the dorsal vertebræ is broad on the more anterior, narrow on the more posterior, of these vertebræ.

[In *B. musculus* the subvertebral ridge is present, but low and blunt on the five anterior dorsal ; reappears better marked on and after the 13th dorsal ; is more decided on the 2nd lumbar, and, after the 2nd, throughout the lumbar region is projecting enough to fill up the concavity of the bodies. Seen from below, the ridges are not flattened out except at their extreme ends.]

7. COSTAL MARKS.—The bodies of the 7th cervical and 1st and 2nd dorsal vertebræ present tubercles on their hinder part. That on the 7th cervical is serial with the inferior transverse processes, those on the 1st and 2nd dorsal are respectively 1 inch and $1\frac{1}{2}$ to 2 inches higher up. They are elongated vertically, their upper ends at about the middle of the body. The posterior slope presents an unfinished surface, as if formerly cartilage-covered, is convex both ways ; in size, about 2 to $2\frac{1}{2}$ inches vertically, $\frac{3}{4}$ to 1 inch in breadth ; that on the 1st is the broader, that on the 2nd the longer vertically ; that on the 7th cervical is a little smaller than the one on the 1st dorsal. The 3rd dorsal, on the corresponding part of the body, has a narrow crescentic mark about $\frac{1}{3}$ inch broad, as if costal ; best marked on the left side. Each of these costal marks is about on a plane with the end of the transverse process of the vertebra behind. As the ribs fall short of reaching the bodies, these marks can have attached only the ligament prolonged from the rib to the body of the vertebra in front.

[In *B. musculus*, costal tubercles exist on the 7th cervical and 1st, 2nd, and 3rd dorsal bodies. They are less elevated, and their surfaces of attachment look more outwards than in Megaptera. The transverse processes here are directed so much forwards, that a ligament passing from near the costo-transverse articulation to the costal tubercle of the vertebra in front would be directed obliquely backwards.]

BODIES OF THE CAUDAL VERTEBRÆ.

8. HÆMAL TUBERCLES, RIDGES, AND GROOVE.—The subvertebral ridge, single and sharp on the 10th lumbar,[1] bifurcates on the 1st caudal. Along the front half of the body of this, the *first caudal*, vertebra there is a shallow groove, bounded by low ridges about 1 inch apart; then the nutritious foramina; and then, along the posterior third of the body, the hæmal tubercles rise rapidly, the space between them $\frac{1}{2}$ inch deep and $1\frac{1}{2}$ inch wide. These tubercles are convex and blunt below, rather than bevelled behind, for the support of the 1st chevron bone. From the 2nd to the 6th the *posterior hæmal tubercles* are larger than those of the 1st, attaining their maximum on the 3rd. They are bluntly triangular on side view. Their anterior border, continuous with the edges of the hæmal groove, is smooth. Their posterior slope presents a flat triangular surface, about 2 inches broad at the summit, against which the anterior slope of the chevron bone rests. Their summits are about 3 inches apart, the triangular space between them 1 to $1\frac{1}{2}$ inch deep. On the anterior half of these bodies (2nd to the 6th) the hæmal groove is bounded by sharp edges.

The *anterior hæmal tubercle*, smaller than the posterior, increases gradually from the 2nd to the 6th, diminishing gradually the interval between the anterior and posterior tubercles. This interval becomes a mere notch on the 6th by the curving forwards of the posterior tubercle. This notch, half-oval in shape and about two fingers'-breadth, is converted on the 7th vertebra into a foramen by the meeting of the posterior and anterior tubercles. The now-constituted *hæmal ridge* is perforated laterally on the 8th as well as on the 7th, forming the first stage of the vertical passage in these two vertebræ. On

---

[1] Rudolphi, *loc. cit.*, speaks of and figures eleven lumbar vertebræ; Van Beneden and Gervais, *loc. cit.*, speak of "neuf lombaires et vingt-deux caudales," but figure the lumbar as ten (pls. x. and xi.). It is evident that the vertebra which I assign as the 1st caudal is really so, and therefore that ten is the correct number of the lumbar vertebræ.

the 8th vertebra, and back to four or five from the end, there is
a continuous hæmal ridge, most projecting at the middle, giving
the vertebra, on side view, a convex outline below.

After the 1st caudal, the *hæmal groove, or fossa,* is concave
longitudinally as well as transversely, owing to the rise of the
tubercles. In width it increases from 2 to 2½ inches at the
middle, where it is widest, and is about ½ inch in depth. Here
two nutritious foramina pierce the bone, about goose-quill size,
one on each side, ¾ to 1 inch apart. On the 7th the hæmal
groove is in length 4 inches; in breadth, in front 3 inches, posteriorly 2 inches; in depth 1½ inches. It now assumes a more
oval form, and on the 10th (the last chevron vertebra) it is in
length 4 inches, in breadth 2¼, in depth 1½; the nutritious foramina close together; the openings of the vertical passages, in
the roof, about 1 inch apart. On the 11th, the long axis of
the oval is still antero-posterior, 2 to 2½ inches long, 1¾ broad,
and ¾ inch deep. The septum between the mouths of the vertical passages is contracted to ⅜ inch. Behind the 11th, the
hæmal fossa becomes oval transversely, and the two great apertures become gradually more separated. On the 12th, the fossa
is broader than long, the great apertures ⅔ inch apart. On the
14th, the length of the fossa is 2 inches, the breadth 2½ inches,
the distance between the two great apertures 1¼ inch. After
the 15th, the fossa becomes broadly diamond-shaped. After the
14th, the distance between the apertures of the vertical passages diminishes actually, but not in proportion to the lessened
size of the bones. At the 18th vertebra they are still ¾ inch
apart, the fossa 2 inches broad, the length ½ inch less. After
the 11th, the two nutritious foramina are transferred from between the great apertures to behind them, and continue so after
the septum between the great apertures has regained breadth.

9. VERTICAL PASSAGE AND FORAMINA.—This large passage,
present in the middle and posterior caudal regions, establishes,
when complete, a communication between the hæmal and neural
canals. In some of the posterior caudal vertebræ it is a simple
vertical canal; in front, as in the 7th caudal (the 12th or 13th
of B. musculus, the 9th of B. borealis), it occurs as a series of
three perforations (4 foramina) in the vertebra, on each side.
On its way from the hæmal to the neural canal three stages

may be recognised—the lower stage, piercing the hæmal ridge; the middle stage, piercing the body internal to the transverse process, showing lower and upper foramen; and the upper stage, piercing the neural arch. The study of the changing position of these great apertures in the transition from the three-perforation condition to the simple condition behind, is the more interesting as they form characters by which these vertebræ of Megaptera may be distinguished from those of B. musculus and B. borealis. These are great apertures, all large enough to admit the point of a finger. They diminish upwards. While within the bone, the passage communicates laterally with the exterior by a system of apertures seen on the side of the bodies.

The *lower stage* exists only on the 7th and 8th, as the perforation in the hæmal ridge, large enough to admit a large finger. Behind the 8th, the passages pierce the roof of the hæmal groove, and become at once concealed in the bone.

The *middle or lateral stage* of the vertical passage exists from the 6th caudal vertebra backwards. On the *fifth*, it has run out into the deep notch bounding the transverse process anteriorly. The deepest part of this notch, on the right side, is marked by traces of the passage, and a wide groove passes up from the notch on both sides. On the *sixth*, it is a canal, $1\frac{1}{2}$ inch in length, in the side of the body, 1 to $1\frac{1}{2}$ inch internal to the transverse process. Between the notch and the foramen there is an interval, 2 inches on the right side, $\frac{3}{4}$ inch less on the left side, and the notch is to the extent of that difference deeper on the left than on the right side. The sixth is the last vertebra with a projecting transverse process. On the *seventh* this stage of the passage is 4 inches in length within the bone, the lower aperture half-way below, the upper aperture less than half-way above the transverse process. From the upper apertures of the 6th and 7th, broad, well-marked grooves are seen to pass up to the neural perforation or notch. These grooves appear to have had a membranous roofing, at least at their lower part. On the *eighth*, the middle stage is now roofed over along nearly the whole side of the body, beginning to be so immediately after the perforation of the hæmal ridge, leaving only a large window between, and reaching up to within 1 inch from the neural perforation. *Behind the eighth*, the passage enters the bone directly from the hæmal groove, and its upper opening is at first high on the side of the bone; then, after the complete neural arch ceases (on the 11th), on what may be termed the upper aspect of the bone; but it is not till the 14th or 15th, with the change to the square form of the body, that the opening is fairly on the top. On the *eight posterior* vertebræ (14th and backwards) it is simply a large, nearly vertical, rounded canal on each side of the flat-sided bodies, through which one may see, and is essentially the same on the three preceding (11th, 12th, and 13th) vertebræ, but their widening body gives the passage a curve outwards. These *eleven posterior* are the vertebræ behind the chevron bones, and those which want a complete neural canal. The lower aperture of the passage is larger than the upper. The latter apertures are wider apart than the lower. The following are the distances between the upper apertures, in inches:—On 18th caudal, 1 inch apart; 14th, $1\frac{3}{4}$; 12th, 3; 11th, $3\frac{1}{4}$; 10th, $3\frac{1}{2}$; 8th, $4\frac{1}{4}$; 7th, $7\frac{1}{4}$; on 6th caudal, $7\frac{3}{4}$ inches apart.

The *upper stage* is present from the 7th back to the 14th. On the first four (7th to 10th), piercing to a completed neural canal; on the second four (11th to 14th), piercing below the low broad ridge which represents a commencing pedicle, to reach the groove or pit which represents the floor of a neural canal. (It is a groove and pit on the 11th, a mere pit on the 12th, 13th, and 14th.) On the second four, the neural perforation does not, as on the first four, open separately on the exterior, but goes in as a narrow passage from the middle stage near the wide upper aperture of the latter. The perforation representing the upper stage is either on one side only, or is not symmetrical. It occurs only on the right side in the 7th, 9th, and 12th; only on the left side in the 8th, 10th and 11th, 13th and 14th. The 7th shows on the left side a groove in the deep notch behind the pedicle, ascending from the second stage of the passage, but not so far forward as the perforation on the right side. On the 6th, the perforation has ceased on both sides, but the notches are not symmetrical, the right being more anterior than the left. Behind the 7th, the perforation is at the middle of the broad neural arch, and on the side on which there is no perforation there is no notch or roundabout groove. On the seven posterior (behind the 14th) the neural perforation ceases. A median septum rises to the level of the edges of the vertebra, separating the funnel-like fossæ into which the passages open, and on each side of the septum nutritious foramina are seen, crow-quill to goose-quill size. In the four previous vertebræ (11th, 12th, 13th, 14th) these nutritious foramina are gathered together into a larger central one in the groove or pit which represents the opened-out neural canal.[1]

Meaning of the Vertical Passage and of its various Conditions.—In endeavouring to find an explanation of the different arrangement of the vertical passage and its foramina along the region, it is observed that the passage is within the body of the vertebra where there are no transverse processes. Also that, with this roofing-over of the passage, the side of the body is nearly filled up, or (as on the 11th, 12th, and 13th) even convex; and that the excavation of the side of the vertebral body begins on the 7th and 6th, as we go forward. If the hinder vertebræ are taken as the type, with the passage concealed in the bone, the commencing excavation on the 7th and 6th would explain the unroofing of the upper and lower parts of the middle stage on these vertebræ and the entire opening up of it on the 5th. Or, going backwards, we perceive that, on the anterior four caudal vertebræ, the segmental blood-vessels do not mark the bone at all, that they begin to groove the bone on the 5th, to pierce it gradually on the 6th and 7th, with disappearing transverse processes, and, after the 7th, to become covered by a bony roof. The adaptation may thus be

---

[1] *Size of the Foramina of the Vertical Passage.*—The size of the apertures of the second stage does not go in proportion to that of the vertebræ. They receive the point of the fore-finger, oval antero-posteriorly, with almost no diminution back to the 10th. The next three are encroached on a little by the broad low pedicles. In the 14th and 15th they have become round and receive the ring-finger above (diameter $\frac{7}{8}$ inch), the fore-finger below. In the 16th they are large enough above to receive the point of the little finger. In the 17th and backwards they are oval transversely. They are always wider below than above. The perforation of the hæmal ridge of the 7th admits the thumb, of the 6th the fore-finger. The perforation of the neural arch of the 7th admits the point of the fore-finger; of the 8th and 9th, the point of the little finger; of the succeeding vertebræ (10th to 14th) less. These are present on one side only, as above noted, but the neighbouring aperture of the second stage is not larger on the side which possesses a neural perforation than on the side which does not.

in affording protection from pressure after transverse processes cease, or protection in the narrow part of the tail. The levelling up, or convexity, of the caudal bodies might thus be regarded as the result of the passages requiring to be roofed over; or the roofing-over may be owing simply to the form of the bodies in adaptation to some other function. The perforation of some of the transverse processes farther forwards in B. musculus and B. borealis is explained by the interruption offered by the great breadth of these processes in them as compared with Megaptera.

10. LATERAL FORAMINA ON THE **BODIES OF THE** CAUDAL VERTEBRÆ.—Besides the usual nutritious foramina—ranged especially towards the fore and back parts of the bodies, the former directed forwards, the latter backwards, here mostly the size of a crow-quill or less,—there is on the caudal bodies a system of larger foramina by which the roofed-over parts of the vertical passage send communications laterally to the surface. Three series may be recognised. They are best understood by beginning behind.

Only the *middle series* are present on the hindmost eleven bodies; on the hindmost eight, as a single aperture at the middle of the side, where the body is constricted, going straight into the vertical passage; on the next three (13th, 12th, 11th caudal) as a pair, arranged antero-posteriorly, the communication having bifurcated as the bodies increased in length and outwardly. This series is continued along the chevron vertebræ close below the line of the transverse processes, on to the 7th. The upper and lower series exist only on the 8th, 9th, and 10th (the three posterior chevron vertebræ). The *upper series*, above the transverse processes, the smallest, occur also as a pair, but are less regular or symmetrical than the middle series. The *lower series*, the largest, three, two, or one in number, pierce the hæmal ridges. That of the 8th is so large that it might be mistaken for the passage, but the passage is continued up inside the bone immediately after perforating the hæmal ridge. The upper and lower series have disappeared on the 7th vertebra from the unroofing of the passage. On the 6th vertebra all the lateral foramina are rendered unnecessary from the shortness of the passage.

Looking to the *direction* of these lateral openings, their general tendency is seen to be downwards, the sharp boundary above, the grooved side below. This is plainly seen in the lower series and on the middle series where they are single. When they become double, in the chevron region, the posterior foramen is directed also backwards, the anterior one forwards. The forward and backward direction is more marked on the upper series. On some a shallow groove is seen to pass between the middle and the upper series. The position of the sharp and the grooved sides of the great foramina of the vertical passage accord with the upward course of the passage.

[*Vertical Passage and Foramina in B. musculus.*—The position of these foramina would be sufficient to distinguish B. musculus from Megaptera. The *lower stage* of the passage is present on the 11th caudal vertebra as a half-oval notch in the hæmal ridge; on the 12th, 13th, and 14th as a foramen. The *middle stage* is present as a foramen in the transverse process from the 7th to the 13th, the three anterior of these in the process proper, the three posterior between the process and the side of the body. In the 7th the foramen is large, 1¾ by 1¾ inch; the bar of bone in front of it 1 inch broad. In the 6th, there is a deep notch in the front of the transverse process instead of a foramen. On the 14th, the middle stage opens high on the body, and behind the 14th it opens at the top. The *upper stage* on the 12th and 13th is present as a notch; on the 14th (the last complete neural arch) as a foramen on the right side, as a notch on the left; on the 15th and 16th, as a foramen on both sides; on the 17th to the 20th, as an open fissure crossing the middle line between the upper apertures of the second stage; on the 21st, 22nd, and 23rd, a septum intercepts this fissure in the middle line.

The great breadth of the transverse processes in B. musculus accounts for their being perforated so far forwards as on seven vertebræ anterior to the first one (the 14th) in which the second stage of the passage is roofed over. The apertures of the passage (judging by the upper apertures of the second stage) are of about the same size and form as in Megaptera, except those in the 8th, 7th, and 6th transverse processes which are larger, the 7th admitting the forefinger, the 6th admitting three fingers.]

PART III.—*continued*.

THE VERTEBRAL COLUMN—*continued*.

NEURAL ARCH AND CANAL.

11. THE LAMINA.—The length of the lamina is remarkable compared with that of B. musculus, but owing to its obliquity (upwards and backwards) and greater thickness the canal is not higher than in B. musculus. The obliquity of the lamina is very slight on the 1st dorsal, increases on the 2nd, 3rd, and 4th, is very marked on the 5th and back to the 1st lumbar; after this it diminishes rapidly along the posterior half of the lumbar region, so that at the 9th or 10th lumbar the lamina is not longer than at the same place in B. musculus. At the last dorsal the length of the lamina proper is about 2 inches, breadth 4, thickness 1½. At the last lumbar the length is 1 inch.

[In B. *musculus* the first three dorsal laminæ are directed a little forwards; obliquity backwards begins on the 5th. The much less length of the laminæ, compared with Megaptera, is recognised on the 7th and 8th, as soon as the articular processes assume the definite

H

quadrate form.   On the last dorsal, the length is 1 inch, breadth 5½, thickness about 1 inch.   The shortness of the lamina and the less width apart of the articular processes, compared with Megaptera, are co-relations.]

12. THE PEDICLE.—The upper limit of the pedicle is defined by a smooth elevation crossing it obliquely from the margin of the articular process to a tubercle on the posterior margin of the pedicle.   This tubercle is present as a flattened pointed projection (1 inch in height at the base, and about ¼ inch in length) from the posterior margin of the pedicle, from the 8th dorsal back to the 4th or 5th lumbar.   On the dorsal vertebræ in front of the 8th there is, however, serial with it, a rough elevation a little in front of the border of the pedicle.   The tubercle appears to be in relation with the anterior inferior corner of the articular process of the vertebra behind it, as if a ligament passed between them, the tubercle a little lower than the process.   It disappears in the lumbar region just where the articular processes cease to have an anterior inferior corner.

[In *B. musculus* there is no projecting tubercle, but along the dorsal region there is a low rough elevation a little in front of the posterior border of the pedicle, at its base and about an inch below the level of the anterior inferior corner of the quadrate articular process behind it.]

The great *thickness* of the pedicle of several of the dorsal vertebræ anterior to the 8th, is owing to its being also the root of the transverse process.   Behind this the thickness does not vary much, diminishing slowly backwards; at the 9th dorsal, 1⅝ inch; at the 13th dorsal, 1⅛; at the 8th lumbar, 1 inch; at the 4th caudal it has increased to 1⅛, and behind this it increases a little.   The *breadth* increases backwards as the thickness diminishes; at the 8th dorsal, 3¼ inches; 10th dorsal, 3½; 1st lumbar, 4; 2nd and 3rd caudal, 4⅛; but after the 3rd caudal it decreases a little.

[In *B. musculus* the *thickness* of the pedicle from the 9th dorsal to the 3rd caudal does not exceed ⅝ inch; from the 13th dorsal to the 14th lumbar it is only ½ inch.   The *breadth* along the lumbar region and posterior half of the dorsal is 4¼ to 4½ inches.]

The *posterior border* of the pedicle is concave, the concavity continuous with that of the lamina, interrupted only where the tubercle above referred to exists. The concavity appears to increase in depth as we go back, owing to the greater projection of the posterior articular process; but it is less in the caudal region, in which, after the 2nd caudal, the pedicle, lamina, and spine all go upwards and backwards with very little obliquity. The *anterior border* rises from the fore part of the body close to the epiphysis, and curves very obliquely upwards and forwards, owing to the form of the articlar process. The anterior inferior angle of the process is 3 inches above the body, giving a high intervertebral foramen compared with B. musculus.

[In *B. musculus* the *posterior* **border** of the pedicle is concave along the whole column. The *anterior border* in the dorsal region curves abruptly forwards to join the lower border of the quadrate articular process, forming nearly a right angle and giving a low intervertebral foramen. The lower border of the process, forming the upper boundary of the foramen, is about 2 inches above the level of the body.]

. The pedicle is set upon the body quite close to the anterior epiphysis all along the column, but at some distance from the posterior epiphysis,—at the 8th dorsal, 1 inch from it; at the 10th lumbar, 1¾ inch. Hence the intervertebral foramen belongs more to the anterior than to the posterior vertebra, although the articular process makes the latter encircle the foramen most.

[In *B. musculus* the pedicle does not arise so near the front of the body as in Megaptera. At the 8th dorsal it is ¾ inch from the anterior epiphysis, 1½ from the posterior epiphysis; at 10th lumbar, the same; at 15th lumbar, ½ inch from the anterior epiphysis, 1½ from the posterior, but curves very rapidly back at its anterior origin.]

Neuro-Central Suture.—Traces of the closed neuro-central suture are present from the 9th lumbar forwards to the 8th dorsal. They are seen and felt as continuous raised lines, along the outside, at about an inch below where the pedicle appears to spring from the body. In front of the pedicle the line would cut off but a small part of the body as belonging to the neural arch. The line is still above the transverse process on the 8th dorsal, and irregular. No definite trace of it can be recognised on the 7th dorsal.

[No certain trace of this suture can be seen in this B. musculus.]

13. NEURAL CANAL.—In *height* the canal shows very little difference along the dorsal region, from $3\frac{5}{8}$ inches at the first to the same at the last dorsal. It is $3\frac{7}{8}$ at the 8th, 9th, and 10th, the highest part of the whole canal. Along the lumbar region it falls from $3\frac{1}{2}$ to 3; along the caudal region from 3 at the 1st to $1\frac{3}{4}$ at the 5th, and to $\frac{3}{4}$ inch at the 10th caudal. The *breadth* diminishes backwards with slight exception, as seen in Table II. At the 10th dorsal, the breadth is but half what it is at the 1st ($7\frac{7}{8}$ inches). From the 10th dorsal ($3\frac{1}{2}$) to the 7th lumbar ($3\frac{3}{8}$) there is very little change.

The floor of the canal presents the longitudinal elevation, with the concavity and foramina on each side of it, noted with the bodies. In the dorsal region the general concavity of the side, of the **Gothic arch** form, is interrupted by an inward bulge opposite where the lamina and pedicle meet. In the lumbar region this minor convexity is higher up. At the 4th caudal the canal assumes an almost square form, changing to a rounded or oval form at the 7th and in the three remaining vertebræ which possess a completed neural arch.

[In *B. musculus* the *height* of the canal increases from the 1st dorsal (3 inches) to the 11th ($4\frac{3}{8}$), and then diminishes (at last dorsal, $3\frac{7}{8}$; at 1st caudal, $3\frac{1}{2}$). The *breadth* diminishes backwards along the dorsal region (1st dorsal, $7\frac{1}{16}$; last dorsal, $3\frac{1}{8}$); increases a very little in the anterior two-thirds of the lumbar region, and again diminishes backwards. Except on the first three dorsal the height of the canal is greater all along than in Megaptera. The breadth is about the same in both from the 3rd dorsal to the 8th lumbar, behind which it is greater in B. musculus than in Megaptera.]

ARTICULAR PROCESSES.

(*Dorsal, Lumbar, and Caudal.*)

14. GREAT ANTERIOR ARTICULAR PROCESSES.—The specially developed articular surfaces of the five or six anterior dorsal vertebræ are seen to be very different in Megaptera and B. musculus when the two series are laid together. In Megaptera

the first is nearly semicircular (2½ by 1¼ inch) and very shallow; ½ inch more in its long direction than that of the 7th cervical and not so flat, and has, like the 7th cervical, a deep pit behind. The 2nd, 3rd, and 4th are increasingly narrower and much bent (the 3rd, length, outwards and forwards, 3 inches, breadth, antero-posteriorly, under 1 inch) and much raised on their oblique outer three-fourths, like the side of a trough. The 5th is a deep ovoid fossa (2 inches by 1½) with a deep non-articular pit (1 inch long) behind it. The 6th (2¼ by 1½) is a shallow mostly vertical surface with a smaller pit behind it. The 7th becomes suddenly lost as an articular cavity, presenting only a smooth apparently non-cartilaginous area bounded by faint ridges.

[In *B. musculus*, the 1st is a semi-elliptical surface (2¾ by under 1 inch) and more raised externally than in Megaptera; the 2nd and 3rd are quite shallow semi-lunes; the 5th, a shallow pointed ovoid (2¼ inches by 1 inch), facing obliquely inward, with a pit behind; the 6th suddenly shows the first internal process, and, in front of this, a flat vertical area without sub-cartilaginous appearance.]

Surveying the series of the ordinary great anterior articular processes, they are strikingly different in Megaptera and B. musculus, especially in the dorsal region; rhomboid and directed upwards and forwards in Megaptera, square-shaped in B. musculus. In Megaptera, back to the 6th lumbar, the concave anterior border of the pedicle rises obliquely to the articular process, meeting it at a rounded obtuse angle, and the anterior border of the process is directed upwards and forwards. The posterior border of the process is oblique, in the same direction. The upper border is a little convex, about ½ inch thick and unfinished. The unfinished edge turns down for ½ inch to 1 inch round the anterior corner, not at all on the posterior corner, so that any further ossification would increase the obliquity of the processes. On the last four lumbar, the angle between the anterior border of the pedicle and the process becomes rounded, and, on and after the 1st caudal, is lost, so that the pedicle and the process now form one interrupted concave anterior border. This concave border becomes less and less oblique, is vertical from the 5th to the 7th caudal, behind which the articular processes are merely low convex ridges. On the last lumbar and two first caudal, the top of the articular processes falls nearly 1 inch compared with those in front; behind this they again rise in height. After the next last lumbar, the posterior border of the process becomes concave backwards, and this with the diminishing obliquity of the anterior border gives the processes, after the 2nd caudal, a straight-up direction, the process increasing somewhat in breadth upwards.

In *thickness* the processes undergo sudden increase at the unfinished top, from $\frac{1}{2}$ inch on the last lumbar to 1 inch on the 1st caudal, and to $1\frac{1}{2}$ inch on the 5th caudal at its middle; after which they diminish, but are still 1 inch thick on the 10th, the last vertebra with a complete neural arch. On the next four vertebræ (11th to 14th) the low ridges, perforated by the passage above noted, represent pedicle and articular processes combined.

The greater *width-apart* of the articular processes in Megaptera, compared with B. musculus, is at once seen in surveying the series from the atlantal or the caudal end. Taking the measurement at the fore part of the processes, inner edge, the distance falls from the 1st to the last dorsal from $11\frac{1}{4}$ inches to $4\frac{1}{4}$; at the 5th and 6th lumbar it is 4 inches; at 9th lumbar, $3\frac{1}{4}$; at 1st caudal, $2\frac{1}{2}$; at 8th caudal, $1\frac{3}{4}$; at the 10th caudal, $1\frac{1}{8}$. The measurements are given in full in Table II. The measurements of a process at the posterior dorsal or anterior lumbar region are—height, $3\frac{3}{4}$ inches; length, 3; thickness, at the unfinished top, $\frac{5}{8}$ inch, at the middle, $\frac{7}{8}$ to 1 inch. The anterior border, and anterior part of the lower border, are much thinner, giving the process a wedge shape.

[*Great Anterior Articular Processes in B. musculus.*

In *B. musculus* the decidedly quadrate form is seen from the 7th dorsal to the 1st lumbar. In front of the 7th dorsal, the upper anterior angle rapidly disappears, reducing the process to a narrow triangular projection, which, on the three first dorsal, is some distance external to the articular surface. The quadrate form, as compared with the form in Megaptera, is obtained by the development of the anterior-inferior and posterior-superior corners. The inferior-anterior forms almost a right angle, but a little rounded off. The measurements at the 14th dorsal are—height, $4\frac{1}{4}$ inches; length, $4\frac{1}{2}$; thickness at the top, $\frac{3}{4}$, at the middle, the same. The thinness of the processes is remarkable compared with those of Megaptera. The development of the upper posterior corner partly carries the process backwards to opposite the fore part of the spinous process, but in Megaptera it is mainly the greater length of the lamina which makes the articular process seem so far in front of the spine. The development of the anterior-inferior corner in B. musculus, bringing the lower border so low as almost to form a right angle with the pedicle, renders the intervertebral notch very low, the anterior part scarcely 2 inches above the level of the body, while in Megaptera the same measurement is about 3 inches.

After the 1st lumbar, the processes become less quadrate, and more like those of Megaptera, the lower anterior corner rounded off, and, after the 5th lumbar, reduced to a mere convexity. After the 14th lumbar, the pedicle and the process form an uninterrupted concave border, going up to the blunt point of a triangular process. The 6th, 7th, and 8th caudal are directed up, with a little concavity on both borders. From the 9th backwards, the short bluntly triangular processes have a direction rather backwards. But all along the lumbar and anterior caudal regions, the posterior angle remains nearly a right angle, and the upper border has very little convexity. In the caudal region, the processes become much thicker, increasing to the 9th, where the thickness is $1\frac{1}{2}$ inches.

In *width-apart*, the distance falls, from the 1st to the last dorsal, from 10 inches to $2\frac{7}{8}$. At the 10th lumbar it has increased to $3\frac{1}{4}$, and from this vertebra, backwards, the width-apart of the articular processes is greater in B. musculus than in Megaptera. (Compare Tables II. and III.)]

15. INTERNAL OR LESSER ANTERIOR ARTICULAR PROCESSES.

—These processes, so well developed in B. musculus, are present here only on the last lumbar and three first caudal vertebræ, and only to a rudimentary extent. On the 2nd caudal, where they are best marked, they project about $\frac{1}{8}$ inch, and have a vertical base of about $1\frac{1}{2}$ inch, the triangular fossa between them $\frac{1}{2}$ inch deep and $\frac{3}{4}$ inch wide in front. On the 3rd caudal they are only low ridges, on the 4th all trace of them is gone, giving a wide uninterrupted space between the great articular processes. On the 1st caudal they are nearly as well developed as on the 2nd, with wider interval between them; on the last lumbar they are less marked, and faint traces of them exist along the trunk vertebræ.

To interpret these processes, begin at the 6th dorsal vertebra, the most posterior of those presenting an articular socket. The outer and inner edges of that socket are represented on the vertebræ behind by two lines, one, now the sharp one (sharp until the internal articular processes appear at the 10th lumbar),

running to the border of the spinous process, the other leaving
that line at about opposite the middle of the great articular
process, and going vertically down; and there is a roughness
where the two lines separate. Arrived at the 9th lumbar, the
vertical line is more distinct, and on the four vertebræ behind
rises into the internal articular process above described. Between these two lines, as we go back along the dorsal and
lumbar regions, the bone is smooth, facet-like, but not as if it
had supported cartilage. The facet, scarcely excavated, is
square-shaped along the dorsal region, is about 2 inches vertically by $1\frac{1}{2}$ inch, and along the lumbar region becomes gradually triangular and smaller. On the 10th lumbar it is in front
of the lesser process and mostly higher up. Where these lesser
processes are present, though in a rudimentary state, the posterior articular processes are undergoing modification, and the
great anterior processes on the 10th lumbar and first two caudal
are lower than on the vertebræ before and behind them.

*The mode of breaking up of the anterior border of the
spinous process* is characteristic in Megaptera. The bifurcation of the very sharp ridge of the border takes place high up.
It is seen to take place a long way behind the articular processes until we go back to the 8th lumbar, when that relation
begins to be reversed. But this is mainly owing to the length
and obliquity of the lamina, which decrease backwards in the
lumbar region. The actual early bifurcation is seen by the
height of the wide triangular concave area from it down to the
level of the roof of the neural canal, $2\frac{1}{2}$ to 3 inches, while in B.
musculus it is scarcely half that height, and is greatly narrower
from the presence of the lower articular processes.

[*Internal Anterior Articular* Processes *in B. musculus.*

In B. musculus these processes are present from the 6th dorsal to the
14th lumbar, with traces farther back. The 5th dorsal is the most posterior of the vertebræ presenting an articular socket. Going backwards, it is seen that the internal, or minor, articular process is serial
with the inner edge of the socket of the 5th dorsal. Increasing
rapidly, it is fully developed from the 10th dorsal to the 10th or
11th lumbar. In its full size it is a triangular flattened process,
projecting $\frac{1}{2}$ to $\frac{5}{8}$ inch with a base $1\frac{1}{2}$ inch in height, pointed and
sharp-edged, and about $\frac{1}{4}$ inch thick at the middle. The median
fossa between the minor processes contracts very much as we go back

(at the 7th, depth $2\frac{1}{2}$ inches, width in front 3 inches; at 10th dorsal, depth 1 inch, width in front $1\frac{1}{2}$) to the 15th dorsal (depth $\frac{3}{4}$, width in front $\frac{3}{4}$); after the 3rd lumbar it begins to widen a little (at the 11th, depth $\frac{7}{8}$, width 1 inch). All along, the fossa between the internal and the great process, about $\frac{1}{2}$ inch deep, will scarcely receive the little finger. The sharp line from the low bifurcation of the anterior border of the spinous process runs to this internal process, from the 6th dorsal backwards.]

16. POSTERIOR ARTICULAR PROCESSES.—After the 5th dorsal there is no posterior articular surface bearing the appearance of having supported cartilage. The process, better developed in the other two finners, is represented by the backward projection where the ridges into which the posterior border of the spine bifurcates join the diverging laminæ, internal to the anterior and upper part of the great anterior processes.

The sloping triangular space included between the bifurcation of the posterior border of the spine and the neural arch, begins higher and is much broader than in the other two finners. In regard to this character Megaptera and B. musculus may be compared at the 4th lumbar, at which the angle formed by the spine is the same in both, and where the neural canal is the same in width ($3\frac{3}{8}$ inches) and only $\frac{1}{4}$ inch lower in Megaptera ($3\frac{1}{4}$) than in B. musculus. In Megaptera the bifurcation begins $6\frac{1}{2}$ inches above the body and $1\frac{1}{2}$ behind the vertical plane of the posterior end of the body; in B. musculus, $5\frac{1}{2}$ above the body and $\frac{1}{2}$ inch behind the same vertical plane. The triangular space in Megaptera is over $1\frac{1}{2}$ inch in breadth at the middle and shallow; in B. musculus about 1 inch and deeper. The contrast increases as we go back, while the triangular space becomes narrower in both. In Megaptera, at the same time, the posterior articular " process " becomes more projecting, back to the 9th lumbar.

On the last lumbar and two first caudal there is a decided triangular mesial projection above the beginning of the bifurcation. It is at this limited region where the minor anterior articular processes exist, and in the fossa between the latter is a low median ridge, not present anywhere else in Megaptera, or at any part in the other two finners. On the last lumbar and three first caudal, where the triangular space between the posterior articular processes runs into the roof of the neural canal, there is a thick prominent mesial ridge about the size of a little finger, but prismatic. It is less marked on the 4th and 5th caudal. Forwards, it is only slightly present in several of the posterior lumbar vertebræ, and re-appears on the posterior half of the dorsal region as a low convexity.

[In *B. musculus* this mesial ridge is very faintly and variably present on some of the lumbar vertebræ.]

The bifurcation of the posterior border of the spinous process is quite peculiar on 3rd, 4th, and 5th dorsal. It begins near the top of the spine and the included space, half-way down, is nearly 1 inch broad, grooved and with a low median ridge. The difference behind the 5th dorsal is owing to the development of this low median ridge as the posterior border of the process, and to the partial filling up of the groove on each side of it.

[In *B. musculus*, the posterior articular processes, back to the 10th dorsal, project far enough to be opposite about the upper and anterior third of the minor anterior processes; more forwards, less as we go back. In the lumbar region the distance between them, anteroposteriorly, increases to about 1 inch. If the posterior processes move straight back, in extension, to that extent, they must go between the minor anterior processes, but they will not touch, as the distance between the latter is considerably greater than that between the two posterior articular processes. The above mentioned character of the 3rd, 4th, and 5th dorsal spines in Megaptera is not seen in B. musculus, in which the posterior edges of all the spines are thin and projecting.]

17. HOMOLOGY AND ADAPTATION OF THE ARTICULAR PROCESSES.—The internal anterior articular processes may be looked on as the true zygomal processes, the great anterior processes rather as metapophyses. Behind the articulation between the 5th and 6th dorsal there is no appearance of the part having been covered by cartilage, but dissection alone can determine that. After that articulation the processes cease to be in contact, the interval increasing from $\frac{1}{2}$ inch rapidly to $\frac{3}{4}$ inch, laterally, as we go back. With this distance between them, laterally, the anterior and posterior processes could not come in contact without more rotation on the axis of the bodies than the intervertebral discs are likely to allow. At the posterior lumbar and the caudal regions they could not even approach each other antero-posteriorly except during severe extension.

[In *B. musculus*, as the posterior articular processes would pass, in extension, internal to the lesser anterior processes, the latter maintain their usual position as the true anterior articular processes. Whether they are serial with the outer or inner margin of the articular surfaces on the anterior dorsal vertebræ does not affect their homology, as they are to be regarded as these processes somewhat rotated, so that the sides of the zygantrum are more vertical.]

TRANSVERSE PROCESSES.

18. GENERAL REMARKS ON THE TRANSVERSE PROCESSES IN CETACEA.—The dorsal transverse processes vary much in the Cetacea, in their place of origin ; in direction, antero-posteriorly or vertically ; in form ; in length, breadth, and thickness, and in the form of their costal facets. Dorso-lumbar transverse processes have generally a forward direction, except some of the posterior dorsal which have a backward direction. This may be said of Mammalia generally, but the greater length of the posterior dorsal transverse processes in Cetacea renders it more evident in them. This is very clearly seen in the toothed Cetacea, *e.g.*, Globicephalus and Phocena communis. The greatly forward direction of the anterior dorsal transverse processes enables the head of the doubly-attached rib to reach the body of the vertebra in front. The more posterior of the ribs have only the costo-transverse articulation, are a-sternal, and have a greater slope downwards and backwards than the anterior, and the transverse processes which support these ribs are directed backwards. The lumbar processes, generally after the first, again assume the forward direction. In the Balænopteræ, the neck and head of the anterior ribs are represented by a ligament, and the very greatly forward slope of the anterior dorsal transverse processes builds up posteriorly the great lateral pyramid formed by them and the cervical transverse processes. There is, in them, the same backward direction of the posterior dorsal transverse processes where the ribs become smaller ; the lumbar, again, after the first, assuming the forward direction.

A typical dorsal transverse process may be divided into two stages, which may be termed the neck and the wing, seen best in Balænoptera musculus. Both borders are at first concave where they join the body. The stages are strongly-marked on the anterior border, where an angle is formed about the middle

by the rapid falling away of the border inwardly on the neck
stage. This angle may be developed into a special projection.
The typical anterior border is thus sigmoid. The posterior
border, at rather internal to the middle, presents a low projection,
sometimes well marked, giving a gently undulating form to the
border. These anterior and posterior intertransverse projections mark off the neck stage, and are apparently the points of
attachment of ligaments or tendons, or the inward part of such
attachments to the more or less expanded wing stage. The
curvatures on the upper and under surfaces of the transverse
processes depend on the amount of upturning of the processes,
and on the amount of thickening at the outer ends to support
the articular surfaces for the ribs. Typical lumbar transverse
processes have the same form except in regard to what depends
on the absence of costal facets and of upturning.

19. THE TRANSVERSE PROCESSES OF MEGAPTERA—DORSAL
REGION.—The most striking characters of the dorsal transverse
processes in Megaptera are their up-curving outwardly, and
their thickness combined with narrowness.

Place of Origin.—The four first spring from the pedicle, the
5th, 6th, and 7th are transition in this respect, the transverse
process gradually springing lower from the common stalk. The
8th may be considered as quite distinct from the pedicle. Behind this, there is more and more separation of the transverse
processes from the pedicles. That this is mainly due to the
processes and pedicles becoming thinner, not to the processes
coming down on the bodies, is seen on a side view. The roots
of the transverse processes are then seen, from the 4th to the
13th, to be on a curved line, convexity upwards. This is in
harmony with the upturning of the processes as an adaptation
to the ribs. The root of the 14th is nearly on a level with those
of the lumbar, being nearly as low as the middle of the body.

The transverse processes, whether united with or separate from
the pedicles, spring from nearer the anterior than the posterior
end of the bodies, from the 1st to the 12th, on which latter the
distances are about equal; but on the 13th and 14th, and on the
1st lumbar, they spring nearer the posterior than the anterior
end. This change does not apply to the pedicles, all of which
spring nearer the front than the back of the bodies.

Direction of the Dorsal Transverse Processes.—The three first have, like the cervical, a downward direction, and are concave below. Above, the two first have little if any concavity. The 1st is not only directed forwards but is bent with the concavity forwards. The 3rd is a little concave above. The 4th is directed a little downwards, and is sigmoid but mostly concave below; above, it is very concave and looks upturned, but is not so high externally as internally. The upturning is very marked from the 5th to the 12th, is but little on the 13th, and the 14th is nearly horizontal. The 5th, as seen from below, ascends 2 inches. Above, it presents a concavity $1\frac{1}{2}$ inch deep, though the outer end is not much higher than the inner, from the connection of the latter with the articular process. The 10th rises 5 inches, measured below; $3\frac{1}{2}$ measured above, the concavity $1\frac{1}{4}$ deep. A line from tip to tip of the transverse processes of the 10th intersects the articular process, cutting off its lower fourth, and passes 3 inches above the level of the body of the vertebra.

The forward direction of the processes is well-marked on the four or five first, and lessens to the 8th, on which the anterior edge of the transverse process is just on a line with the front of the body. The distances to which the transverse process reaches beyond the front of the body of the vertebra are—the 1st, 2 inches; the 3rd, 3 inches (angle 22°); the 5th, $2\frac{3}{4}$ inches; the 7th, $\frac{3}{4}$ inch. The backward direction of the processes begins with the 9th, increases to the 12th or 13th (angle about 15°), and diminishes on the 14th. The hinder edge of the process begins to take the backward direction on the 7th. The most posterior part on the 9th is flush with the back of the body. The distances to which the processes pass behind the plane of the back of the body of their vertebra are—the 11th, 2 inches; the 13th, 2 inches; the 14th, $1\frac{1}{4}$ inch.

In *form*, the dorsal transverse processes pass through a transition from the form of the cervical to that of the lumbar,—the anterior, flattened with the surfaces before and behind, prismatic in transition, and, towards the lumbar region, flattened with the surfaces above and below. The great convexity which the 2nd has gained on its posterior surface becomes developed into and remains as the posterior border as we go backwards. The superior border remains as the anterior border. The inferior border is gradually carried backwards to below the flattening process, and forms the lower edge of the prism; is thick internally, is continued externally in front of the fossa, as the lower edge of the prism, and disappears as we go back, except as the anterior boundary of the costal fossa. The transition is rapid on the 4th; the 14th is about as flat below as it is above, like a lumbar process, though not so thin.

The borders on the first five are, the anterior, like the posterior cervical, concave; the posterior, convex. The 6th is transition in these respects. On the 7th, and backwards, these curvatures appear reversed, but anterior and posterior intertransverse projections make their appearance on the 6th, and continue with variable prominence after the 6th, so that the true form of the hinder edge is that of a low prominence with gentle concavity on either side, and that of the anterior edge sigmoid, the concavity on the neck stage.

On their upper surface the transverse processes are, after the first four, concave in their whole length; from the 5th to the 12th very much so, owing to the great rise of these processes outwardly. On the 13th, the concavity is much less; the 14th shows a shallow sigmoid curvature, the concavity on the outer half.

Length, Breadth, and Thickness of the Dorsal Transverse Processes.—In *length* they increase gradually from the 1st to the 14th, and begin to diminish from the 4th lumbar backwards. In *breadth* they increase from the 1st to the 10th, and diminish a little on the 13th and 14th. The wing stage is marked off by the low intertransverse projections and by the greater breadth, but the difference in breadth is not very great. On the 7th the neck is 3 inches in breadth, the wing 4½. The spaces between the processes, in the articulated skeleton, are about as wide as the processes themselves, a good deal wider internally, scarcely so wide externally. In *thickness*, taken at the middle, the 3rd, 4th, and 5th processes are the greatest (2½ inches), after which the thickness diminishes, and rapidly so on the 13th and 14th.

20. COSTAL FOSSÆ ON THE TRANSVERSE PROCESSES.—The fossæ on the transverse processes for articulation with the ribs

are best marked on the 10th and 11th. They are shallow cavities, in form the lower half of an ovoid, the point directed inwards and downwards. The 10th measures 4 inches in the inward and downward direction, $3\frac{1}{4}$ inches antero-posteriorly; the greatest depth is $\frac{5}{8}$ inch. On the three first there is no pit, only the blunt convex unfinished end of the processes. The thick blunt end of the 3rd is more sloping below than above, but there is no hollow or costal mark. The fossæ begin on the 4th, where it is well marked, increase in size and in depth backwards to the 10th, and diminish in size, but have sharper edges, on the 11th and 12th. On the 13th there is a very shallow triangular facet, 2 inches wide. The 14th process presents only an elliptical outer end, twice the thickness of the end of the 1st lumbar, the lower and back part of which shows a slight costal bevelling.

The fossæ reach to the outer end of the process and occupy the posterior of the two under surfaces, as if formed by a bifurcation of the inferior border of the process, but the anterior edge of the fossa is the prominent one. The surface in front of the fossa, at first narrow, becomes gradually broader as we go back, attaining on the 12th the same breadth as the fossa. The fossæ look mainly downwards, to a less extent outwards and backwards. Viewed from the side the ends of the processes are seen to form crescents over the fossæ, directed very obliquely forwards and downwards, with a short continuation forwards from the 7th to the 12th. These margins and the fossæ would probably be more sharply marked in the completely ossified state. Here, taking the 7th, the thickness of the unfinished edge, opposite the middle of the fossa, is 1 inch.

21. LUMBO-CAUDAL TRANSVERSE PROCESSES.—In *place of origin* all the lumbo-caudal transverse processes, after the first, are on a line with about the middle of the bodies. After the 2nd lumbar they spring from rather nearer the front than the back of the bodies; in the caudal region it is rather the opposite, this arising from the increased depth in the caudal region of the anterior concavity of the neck of the process.

In *direction*, the lumbar transverse processes have a more or less forward tendency after the 1st, which is directed slightly backwards (angle 5°). The 2nd is directed a little forwards, the

3rd and 4th more so, the 5th to the 8th decidedly forwards (angle at the 7th, about 22°), the 9th a little, the 10th very little. The direction of the caudal transverse processes is variable, depending on the non-development of the anterior or posterior angles of the wing, external to the intertransverse projections. The direction of the 1st is straight out; that of the 2nd and 3rd backwards, owing to the failure of the anterior angle of the wing; that of the 4th a little forwards, that of the 5th and 6th very much forwards, owing to the want of the posterior angle and of almost the whole of the wing, and to the great depth of the anterior concavity of the neck. The 7th and 8th are mere lateral ridges. The change of the posterior border to the backward direction begins on the 9th lumbar and continues to the 4th caudal.

In *length* the processes continue the same as the last dorsal to the 3rd or 4th lumbar, after which they diminish; after the 6th lumbar rapidly, after the 3rd caudal very rapidly. On the 6th caudal they are very short, on the 7th and 8th they are represented by a mere tubercular trace. In *breadth* they increase a little by expansion of the outer half, to the 8th lumbar, after which the breadth diminishes as we go back. In *thickness*, taken at the middle, they are pretty uniform. On an average the thickness is about 1 inch, about half that of the dorsal transverse processes.

In *form* the processes show more than in the dorsal region the distinction between neck and wing, as they broaden outwardly and shorten, on the posterior lumbar and anterior caudal region. On the anterior border, the intertransverse projection becomes more marked on and after the 9th lumbar, from the falling away of the wing external to it. This becomes more marked on and after the 1st caudal, giving the anterior border a backward direction external to the projection. Thus the typical sigmoid form of the border becomes more marked backwards. From the 9th lumbar backwards, the concavity is more striking, partly from being deeper, partly from now occupying a greater proportion of the length of the now shortening processes. From the 3rd to the 6th caudal, the concavity becomes increasingly deep and narrow; and on the 5th is manifestly the opening out of the vertical passage seen on the vertebræ behind it. On the posterior border, the intertransverse projection is seen in various degrees, giving the wavy outline.

The ends are all convex and incompletely ossified, except the 5th and 6th caudal, which are sharply finished. The unfinished state being only at the ends seems to indicate that further ossification would have added to the length but scarcely to the breadth of the processes. On the whole, these processes do not expand much, although on the last five lumbar it is noticeable enough. The breadths of the narrowest and broadest parts of the process, respectively, are—of the 1st lumbar, $3\frac{1}{4}$ and 4 inches; of the 8th, which has the greatest expansion, $3\frac{1}{2}$ and $5\frac{1}{2}$; of the 10th, $3\frac{2}{3}$ and $4\frac{2}{3}$.

Comparing the breadth of the lumbar processes at their outer half with that of the spaces between them, the first two spaces are rather less than the processes; the five next about equal. From the 8th vertebra backwards, the spaces become wider and wider compared with the processes. Between the first five caudal the spaces are two to three times as wide as the processes. This depends not only on the processes becoming narrower, but on the increased distance in the caudal region between the bodies of the vertebræ.

22. GENERAL SURVEY OF THE TRANSVERSE PROCESSES IN MEGAPTERA.—Viewed from the caudal end and from above, the extreme outline of the processes, from the last costal vertebra backwards, has the figure of the hinder half of an ellipse, the convex edges tapering very rapidly on the 4th, 5th, and 6th caudal, the latter being the first which shows a prominent transverse process. This view of Megaptera is remarkable for the absence of transverse processes on the fifteen posterior caudal vertebræ, a space of about $7\frac{1}{2}$ feet, giving the caudal vertebræ a clipped appearance. Forwards, what strikes the eye is the upturning of the dorsal transverse processes, becoming very marked on the 12th. At the same time the outline figure narrows forwards to form the fore part of the ellipse, but the tilting up of the dorsal processes breaks the outline. Viewed from the atlantal end, the upturning of the dorsal processes is still more apparent, rising like the ribs of a ship. The outline along the dorsal processes shows contraction forwards, but not much convexity. To the eye the 14th dorsal seems the widest,

I

with diminution immediately behind and before. The eye does
not detect that the first three lumbar are equally wide, nor
would the eye detect so great a diminution as 6 inches from
the 14th to the 8th dorsal.

[23. TRANSVERSE PROCESSES IN B. MUSCULUS COMPARED WITH THOSE
OF MEGAPTERA.

The transverse processes differ much from those of Megaptera,
in expansion of their outer two-thirds, in being much thinner, and in
being very little turned upwards in the dorsal region.

24. DORSAL REGION.—*Origin.*—The root common to the pedicle
and the transverse process on the first four is more oblique than in
Megaptera. On the 7th, the pedicle and the process are as distinct
as on the 8th of Megaptera. Behind this, as seen from above, the
processes spring from lower on the bodies than in Megaptera. This
is owing to their thinness. Seen from below, they do not spring so high
from the bodies as in Megaptera, but from the 4th to the 13th their line
of origin is slightly convex upwards. At the 15th, the process is thin
enough to have fallen quite to the level of the middle of the body.

Direction.—The first is straight. The three first are decidedly
concave above. Viewed from above, the processes appear to the eye
to rise outwardly from the 5th to the 12th, but it is from the 7th to
the 13th that the outer end of the process is really higher than the
inner part, the 13th very little. The rise above the costal fossa
renders the outer half concave. From the 6th, diminishing backwards, the inner half of the process is rather convex. A line from
tip to tip of the transverse process of the 10th shows a rise of $1\frac{1}{2}$
inch from the neck to the tip of the process. The line is on a level
with the body at its hinder edge and $1\frac{1}{2}$ inch below the articular
process. The concavity is about $\frac{1}{4}$ inch deep. On the 7th it is twice
that, the depth increasing forwards owing to the inward rise of the
process to the articular process. Viewed from below, the processes
rise a little outwardly from the 7th to the 13th. The *forward direction* of the processes is greater than in Megaptera. The distances to
which the processes pass in front of the plane of the body of their
vertebra are—the 1st, $4\frac{1}{2}$ inches; the 3rd, $5\frac{1}{2}$; the 5th, $3\frac{1}{2}$ (angle
40°); the 7th, $1\frac{1}{2}$; the 9th, 1 inch. The 10th is nearly straight out.
The backward direction begins on the 11th and 12th. The backward direction of the posterior border begins on the 8th. On the
11th, the most prominent parts of the process are, respectively, $\frac{1}{4}$ inch
behind the front of the body and $\frac{1}{2}$ inch behind the back of the body.
Behind this, the backward direction of the processes is much less
than in Megaptera. The posterior part of the process is 1 inch
behind the body on the 13th (angle 5°); on the 14th, $\frac{3}{4}$ inch; on
the 15th, it is $\frac{3}{4}$ inch in front of the back of the body.

Form.—The prismatic form is well marked on the 3rd and 4th,
on their inner half, and throughout on the 5th and 6th, especially
at their outer part. Behind the 6th the processes grow thinner and
thinner by the gradual disappearance of the lower edge of the prism.

Seen from below, the processes, from the 2nd backwards, are concave as far out as the beginning of the costal fossa, though very shallow after the 9th. In Megaptera, after the 5th, the processes appear convex below from the upturning of the process and the far in position of the costal fossa. The form of the processes is further noted with the breadth.

Length, Breadth, and Thickness of the Dorsal Transverse Processes.—
In *length*, the processes increase to the 13th, and begin to diminish after the 14th. In *thickness* (vertical measurement, taken at the middle), the 5th is much the greatest, behind which there is rapid diminution. In *breadth*, which is their most remarkable character, they increase backwards to the 11th ($7\frac{1}{2}$ inches), and then diminish backwards. The expansion is rapid on the 6th (breadth, 6 inches). From the 7th to the 14th the difference is not great. From the 9th to the 13th there is almost no difference in the breadth of the neck of the process (about $4\frac{1}{2}$ inches). On the 11th the wing is within about an eighth part of being twice as broad as the neck; and from the 7th to the 11th the breadth of the wing is within $\frac{1}{4}$ inch of being as great as the length of the body. The anterior intertransverse projection, or angle, marking off the two stages of the process, is about the middle of the border. The projection on the posterior border is internal to the middle, is well enough seen on most of the processes, and gives the undulatory form to this border. The anterior half of the wing is thin. The beam of the process runs out from the neck to opposite the costal fossa, which it supports. The expansion on one side of the process tends to alter its direction, compared with that of a process which has the beam only. In the articulated skeleton, the spaces between the processes, from the 6th backwards, are, between the necks, about equal to the breadth of the neck; between the wings they average about $1\frac{1}{2}$ inch, being less than a fourth part of the breadth of the processes. That this remarkable expansion of the outer half of the dorsal transverse processes in B. musculus, compared with Megaptera, is not merely a distinction of age is evident from the fact that, while in this B. musculus the anterior border of the wing has the appearance of a part from which a strip of cartilage has been removed, the anterior border in Megaptera has already a finished appearance.

*Costal Fossæ.—*The costal fossæ are very different from those of Megaptera. They are most characteristic from the 6th to the 10th. They are placed on about the posterior half of the broad end of the process, rather than below it, and are elongated antero-posteriorly. (On the 7th, length, 4 inches; vertically 3 inches; depth, $\frac{3}{4}$ inch.) A rounded-off angle below gives the fossa a low triangular form, changing to the elliptical as we go back from the 9th and 10th. The overhanging arched upper boundary has very little of the obliquity seen in Megaptera. The whole outer margin of the broad wing has a gently sigmoid form, the posterior half, or a little more, above the costal fossa. The fossæ look mainly outwards, with a lesser degree of direction backwards and downwards. The outer margin of the processes, though not fully ossified, are much sharper than in Megaptera. On the 7th, over the middle of the costal fossa, the margin is ¼ inch thick. On the 12th, 13th, and 14th they are narrow elliptical facets, very shallow; on the 13th and 14th quite on the end of the process and looking straight outwards. The transverse process of the 15th bears no mark for its rib, and is very little thicker than that of the first lumbar. The anterior half or two-thirds of the fossæ have been covered with cartilage; when the 12th is reached the whole area has been so. This contrasts with Megaptera in which the fossæ do not present the appearance of having been covered by cartilage.

25. LUMBO-CAUDAL TRANSVERSE PROCESSES IN B. MUSCULUS.—*Origin.* —Like the hinder dorsal, they are on a line with about the middle of the bodies. They spring nearer the front than the back of the bodies, a little on the 1st lumbar, increasing in this respect back to the 11th; less so on the three next; on the 15th rather the reverse; from the 1st caudal to the 6th, decidedly nearer the back of the body, this being due to the greatly increased depth of the anterior concavity after the vertical foramen has opened into it. After the 7th caudal, in which that foramen pierces the process, the process springs much nearer the front than the back of the body.

The forward *direction* begins with the 3rd lumbar, and increases to the 7th. From the 10th the forward direction lessens, and at the 15th is very slight. On the first six caudal the processes are on the whole horizontal, varying a little according as the anterior or posterior corners are most developed. Behind the 6th the direction of the processes is noted with their form. In *length*, they diminish a little from the 1st to the 10th lumbar, and from the 11th rapidly backwards. The thickness is about half that of the dorsal processes. In *breadth*, they vary little from the 1st lumbar to the 3rd caudal, averaging about 6½ inches; the 13th and 15th lumbar have the exceptional breadth of 7 inches. The regained breadth of the 7th caudal, and of the three or four behind it, is owing to the change of form consequent on the vertical foramen.

Form.—The lumbo-caudal processes, back to the 6th caudal, present the same flattened and outwardly expanded form as the posterior dorsal processes, but are thinner all along the process. Back to the 12th lumbar the wing is less square-shaped, as the expansion is less and also more gradual. The distinction between neck and wing is less marked, but the anterior intertransverse projection is recognisable, and the projection on the posterior border is rather better marked than on the dorsal processes. Behind the 12th, the lumbar processes become more convex externally from the rounding off of the angles. The first six caudal show this degeneration to a greater extent; the neck shortens, and the wing-stage is represented only by a semicircular end beyond the anterior and posterior intertransverse projections. The anterior concavity of the neck deepens, and on the 7th caudal, is enclosed, forming the foramen of the vertical passage. The stunted processes behind this, having the foramina, are very broad, springing now from as far forward as where the body joins its epiphysis. Behind the 8th, however, the foramen is at or behind the middle of the process.

General Survey of the Transverse Processes in B. musculus.

Viewed from the caudal end and from above, the general outline of the transverse processes shows **but little** of that convexity presented by the processes of the lumbar and anterior caudal regions of Megaptera, or by the whole sweep of the outline in B. borealis from the neck to nearly the middle of the caudal region. In this B. musculus the outline is nearly flat from about the 12th dorsal back to nearly the end of the long lumbar region ; and along the caudal region the outline tapers gradually backwards, with very little convexity. If compared to an ellipse, it would be the hinder half of a very narrow ellipse, beginning at the 13th dorsal vertebra. Only the 13 or 14 posterior caudal vertebræ appear destitute of transverse processes, giving a length of 4½ to 5½ feet, as compared with the 7½ feet of Megaptera, without transverse processes. Viewed from the atlantal end, the eye would take the 8th dorsal for the widest, and recognises very little diminution on the 7th and 6th. The dorsal region as a whole shows a figure bounded by a gently convex outline. Viewed from above or below, the great breadth of the transverse processes contrasts strongly with their little expansion in Megaptera. **The** difference is least in **the posterior** lumbar region in Megaptera, **but is** great throughout.]

SPINOUS PROCESSES.

26. FORM.—The most striking character of the spines in Megaptera is their narrowness and but little of that expansion at the end which B. musculus shows to so great an extent. This may be in part owing to incomplete ossification in this Megaptera, but the exposed surface from which the cartilage has been removed belongs to the top with a tapering prolongation below the posterior angle for about 1 inch only. Further ossification would increase the length, but it could not well give a terminal expansion so great as in B. musculus, so as to bring the expanded ends nearly in contact, or actually so, for from 3 to 5 inches down from the top. Taking the last dorsal vertebra, the breadths at the middle and at the end are—in Megaptera, 4¼ and 5½ inches; in B. musculus, 5¼ and 8½. **In the** articulated skeleton the intervals between the end of the dorsal processes are fully 2 inches, while in B. musculus the expanded **ends may be** in **contact or** about 1 inch **apart.** Rudolphi's figure of the skeleton **of** Megaptera shows the **spines** almost square-topped, **and not** broader **at** the top **than at the** middle. In the **figure of the** skeleton by Van Beneden and Gervais (pls. x. and xi.) the tops of the spines are less rounded, but not more expanded than in this Megaptera. Their figure of

B. musculus (pls. xii. and xiii.) does not show the great expansion
of the ends of the spines presented by this B. musculus.

27. LENGTH, BREADTH, AND THICKNESS OF THE SPINOUS
PROCESSES.—In *length* the processes increase from the 1st dorsal ($3\frac{1}{2}$ inches) to the 5th lumbar (the 5th, 6th, and 7th lumbar
each 14 inches), and decrease from the 8th lumbar to the 10th
caudal ($\frac{3}{4}$ inch). The longest transverse processes are the six
before the above group of longest spinous processes.

The *breadth* of the processes at their middle does not vary
much from the middle of the dorsal region to the earlier part of
the caudal, the broadest being the last lumbar and 1st caudal
($5\frac{1}{2}$ inches). The expansion at the top follows the same order,
about $\frac{1}{2}$ inch to 1 inch broader than at the middle. The 8th
and 9th dorsal come in exceptionally as broad at the top (6
inches) and are nearly as broad at the middle ($5\frac{1}{4}$ inches) as the
last lumbar and 1st caudal.

In *thickness* the spines increase backwards to the 1st lumbar,
and decrease backwards from the 2nd, but the difference is not
great. From the 4th to the 7th dorsal it is 1 inch; on the
next three, it is $1\frac{1}{8}$ inch; on the remaining four dorsal, $1\frac{1}{4}$ inch;
on the 1st and 2nd lumbar, $1\frac{3}{8}$ inch; on the 3rd to the 8th
lumbar, $1\frac{1}{2}$; on the 3rd and 4th caudal it is again 1 inch.

The thicker part of the process is generally behind the middle.
This is seen on the top by the generally greater thickness of the
posterior third of the border, giving it a narrow ovoid form.
This *ovoid upper margin* is seen from the 8th dorsal to the
8th lumbar (10th dorsal, at middle 1 inch, behind $1\frac{1}{8}$; 14th
dorsal, $1\frac{1}{4}$ and $1\frac{1}{2}$), is most marked on the 1st lumbar ($1\frac{1}{4}$ and
$1\frac{5}{8}$), and then diminishes backwards. From the 9th lumbar
to the 2nd caudal it is absent, and is again seen to a moderate
extent on the spines behind that. On the last dorsal and two
first lumbar, the thick part of the ovoid end is thicker (by $\frac{2}{8}$ inch)
than the thickest part of the middle of the shaft; but, taken at
the middle of the ovoid, so as to give an average thickness, the
end of the spinous process is generally thinner than the shaft
by from $\frac{1}{8}$ to $\frac{3}{8}$ inch.

28. DIRECTION OF THE SPINOUS PROCESSES.—The spines of
the 1st, 2nd, and 3rd dorsal are directed backwards, owing to the
greater slope of their anterior border. The 4th, at first some-

what triangular, is directed straight up; the 5th, now square-shaped, a little forwards; the 6th, straight up; the 7th, a little backwards. The remaining spines all slope backwards, increasingly so to the 14th dorsal. After the 1st lumbar the slope diminishes a very little to the 6th, then a little more to the 9th, and still more from the 10th lumbar to the 3rd caudal. Behind the latter the slope begins to increase, but the processes are too irregularly formed to give precise indication.

A convenient way of measuring the amount of the slope is to take a perpendicular from the plane of the back of the body. But the result by this method is influenced by the length of the spine, and in Megaptera (as compared with B. musculus) by the length and backward obliquity of the laminæ in the dorsal and anterior lumbar regions, thus carrying the base of the spines backwards in these regions as compared with the middle and posterior lumbar regions. The result by the perpendicular line, therefore, does not always correspond to the real obliquity, as ascertained by taking the angle.

The figures in the first column of the subjoined table (Table IV.) give the distance of the *back* of the body behind the *middle* of the top of the spine; the second column gives the true obliquity, the angle formed by the axis of the body and the axis of the spinous process. For comparison, the same measurements in B. musculus are given.

TABLE IV., *giving the Amount of Obliquity of the Spinous Processes in Megaptera and B. musculus.*

Vertebra.	Megaptera.	Angle.	B. musculus.	Angle.
1st dorsal,	1 inch behind,	75°	2 inches in front,	100°
6th ,,	1 inch in front,	90°	2 ,, ,,	90°
9th ,,	Same plane,	80°	1½ ,, ,,	88°
11th ,,	1 inch behind,	65°	Same plane,	70°
14th ,,	3½ ,, ,, (greatest),	55°	2 inches behind,	65°
3rd lumbar,	2½ ,, ,,	56°	4 ,, ,,	57°
6th ,,	1½ ,, ,,	57°	4½ ,, ,,	54°
10th ,,	Same plane,	65°	5½ ,, ,,	50°
11th ,,	,, ,,	...	6½ ,, ,, (greatest),	46°
1st caudal,	½ inch in front,	68°	2½ ,, ,,	55°
3rd ,,	1 ,, ,,	70°	½ ,, in front	60°
6th ,,	2 ,, ,,	60°		55°

Besides the points above referred to, the following may be noted of the *form* of the spinous processes. *Anterior border,* on the first three dorsal, sloping and convex, like the 7th cervical; from the 4th dorsal backwards, it has a slight general concavity, but scarcely so on the 6th and 7th; 4th caudal, exceptional from want of development of the anterior corner. *Posterior border*, from the second dorsal backwards, a slight general concavity, but from the **6th** dorsal to several of the anterior lumbar very slight. The posterior border of these spines **might** at first appear as if having a general convexity were the eye to take in the upper part of the lamina, so little do the posterior articular projections form a distinct angle till we get well back into the lumbar region. On side view, the *ends* are pointed on the first three dorsal, nearly flat from the 4th to the 11th dorsal, more rounded back to the 6th caudal, after which they become less rounded. The spines of the 6th and 7th caudal assume a triangular form, base above, from the prolongation of their posterior corner. On the 8th, 9th, and 10th the spines are merely uniform decreasing ridges.

[29. Spinous Processes in B. musculus.—The great expansion of their ends is already referred to. In the dorsal and lumbar regions, beginning at the 5th dorsal, the expanded ends are pretty close, 1 inch, more or less, apart, as now articulated; and from the last dorsal to the 12th lumbar, they bear marks of having been in contact. The now rough *facets of contact* begin from 2 to 4 inches down from the top of the spine behind, are mostly 3 inches in length, and ½ inch in breadth, the border being specially thickened to form them. They are particularly marked on each side of the 7th, and were noticed there in the dissection to have a soft covering. These surfaces are distinct from the unfinished edges, which, however, go down below both corners twice as far as in Megaptera, thus making even further anterior-posterior expansion possible.

From the 10th dorsal to the 12th lumbar, the expansion is on the posterior part of the ends, very little on the anterior part; but on and after the 13th lumbar, the anterior corner also expands, and, on the 2nd, 3rd, and 4th caudal, is more expanded than the posterior corner. The very marked expansion of the posterior part of the end, as seen especially from the 13th dorsal to the 12th lumbar, is downwards as well as backwards, so that the processes come in contact some way down on the one behind. This is a striking character of B. musculus as compared with Megaptera.

Another contrast between Megaptera and B. musculus is the expansion, likewise backward, of the spinous process below, connected with the better development of the posterior articular processes in B. musculus. By this expansion the spinous process in B. musculus gains a breadth on an average of about 1 inch, or more. It is well seen from the middle of the dorsal region back to the 1st caudal, rather increasing as we go back to near the end of the lumbar region. This prominence renders the posterior border of the spines very concave. In Megaptera it is pronounced only on the last lumbar and 1st caudal; back at least to the last few lumbar it is so low as to be only enough to give a convex outline to the spine and lamina, viewed together.

Length, Breadth, and Thickness of the Spines.—In *length* they increase from the 1st dorsal (3¼ inches) to the 10th lumbar (18¼, the 9th and 11th each 18 inches), and then decrease backwards to the 1st caudal (13¾), finally ceasing on the 14th caudal. In *breadth*, at the top, they increase backwards to the 14th dorsal (8⅛ inches, breadth at middle, 5½), and decrease from the 15th, but there is not much diminution till after the lumbar region (1st caudal, 7¼, breadth at middle, 5). In breadth at the middle, the difference from the 7th dorsal (5¼ inches, breadth at top, 6¼) to the 1st caudal is not great; the broadest, the 9th, 10th, and 11th, are 6 inches, with breadth at the top of 7½ inches. In *thickness*, they are considerably less than in Megaptera, although in B. musculus the processes are longer and broader. At the middle they increase in thickness from ⅜ inch on the 1st dorsal to ⅝ on the 6th dorsal, to ⅝ on the 10th dorsal; continue pretty steady at ⅝ or ⅞ back to the 2nd caudal, after which they decrease to ½ inch to the 7th caudal.

The narrow *ovoid upper border* is seen from the 8th dorsal to the 13th lumbar (10th dorsal, at middle ⅝, on posterior third ⅝; 14th dorsal, ⅝ and 1; 1st lumbar, ⅝ and 1¼; 11th lumbar, ⅝ and 1½; no other exceeds 1¼). On the 14th lumbar the end is contracted at the middle, and on the last lumbar and first four caudal the ovoid is reversed, the broader end forwards, ⅛ inch broader than at the middle. Comparing the thickness of the ends with that of the middle of the shafts, the thickness is the same in front of the 9th dorsal and behind the 5th caudal. From the 8th dorsal to the 13th lumbar, the thick part of the ovoid is thicker all along than the shaft; but the middle of the ovoid is thinner than the shaft by from ⅛ to ⅜ inch. The ends are more finished here than in Megaptera. With thinner processes, the increased thickness on the posterior third of the end is relatively greater than in Megaptera.

Direction.—The spinous processes differ from those of Megaptera in direction. Those of the 1st and 2nd dorsal are directed forwards. The backward slope begins in both on the 7th, is less in the dorsal region, and greater in the lumbar region, than in Megaptera. While in Megaptera the slope begins to diminish at the 14th dorsal, in B. musculus it goes on increasing back to the 11th lumbar. The measurements for this comparison are given above in Table IV.

The spinous processes of *B. borealis* are very different from those both of Megaptera and of B. musculus. The differences, too numerous to be described here, will be given afterwards in a comparison of B. musculus and B. borealis. The spinous processes of these three whales may be characterised generally, in Megaptera as rhomboid ; in B. musculus as of battledoor shape ; in B. borealis, the dorsal as parallelograms, the lumbar as of hour-glass shape.]

PART III.—*continued.*

CERVICAL VERTEBRÆ.

The measurements in the following Table may be compared with those given in the Table of Measurements of three series of the cervical vertebræ of B. musculus, in this *Journal*, vol. vii., November 1872.[1]

[1] The paper referred to, "On the Cervical Vertebræ and their Articulations in Fin-Whales," contains a full account of the characters of the cervical vertebræ in Balænoptera musculus and in B. rostrata, and of their articular surfaces and ligaments, including the variations presented by the three complete sets of the cervical vertebræ of B. musculus and by other specimens of atlas and axis of that species. The characters in B. musculus there described are assumed here, and are occasionally referred to only for comparison with those of Megaptera. It will be understood that "the B. musculus" noted in the following account is the 50-feet-long B. musculus with which the Megaptera is compared throughout this paper, unless the other specimens described in the previous papers are mentioned.

30. TABLE V.—*Measurements of the Transverse Processes of the Cervical Vertebræ, in addition to those contained in Tables II. and III., given in inches.*

	Megaptera.							B. musculus, 50 feet long.						
	Atlas.	Axis.	3	4	5	6	7	Atlas.	Axis.	3	4	5	6	7
1. Length of upper, . .	3⅞	6½	6	6½	6½	6⅓	6	6¼	10¾	9¼	9	9	8½	8¼
2. Length of lower, . .	...	5¼	3¼	2¾	¼	...	...	...	10	7½	7½	8¼	3¾	...
3. Breadth of plate beyond the rings.	...	...	...	...	...	...	...	...	5	...	...	...	...	...
4. Transverse diameter of the rings.	...	...	...	...	...	...	...	...	5¼	7	7	6¾	...	...
5. Greatest distance between upper and lower processes.	...	3⅝	5¼	6	...	...	...	...	3¼	6	6¼	6⅝	6¾	...
6. Distance between their ends.	...	3	5⅝	6	...	...	...	...	...	1¼	1½	¼	5¼	...
7. Weight in ounces,[1] .	209	176	...	54	...	...	73	176	212	...	79	...	...	108

31. THE ATLAS.—*Anterior Aspect.*—The groove between the condyloid cavities is wide. This might seem to be a distinctive character in contrast with the narrowness of the groove in B. musculus, as seen in four of my specimens. In two of them it is reduced to the condition of a mere median furrow. But in the 50-feet-long B. musculus the width is not much less than in the Megaptera. In the latter the width is—below, which is also the narrowest part, 1½ inch; at middle, 1¾; above, 2⅝. In the B. musculus, the corresponding measurements are, 1½, 1½, and 2⅛ inches; the narrowest part, towards the lower end, 1¼. The widest in the other four specimens of B. musculus is ⅝ inch, and the groove has that width nearly all along. In both the Megaptera and the 50-feet-long B. musculus the furrow for the capsular ligament of each condyle is seen at the edge of the wide groove. In the dissection (1872, *loc. cit.*, p. 14) I found the median interval so narrow that the two capsular ligaments seemed to have coalesced, and the median septum thus formed to be disappearing, but I could not be quite sure that the seeming partial disappearance of the median septum was not due to giving way of the parts. The narrowing of the space

[1] The weight of the first caudal vertebra is, in Megaptera, 304 ounces, in the B. musculus, 368 ounces.

between the condyloid cups, as seen in these five specimens of B. musculus, may be a matter of age, but in the atlas of a sixth great finner (referred to, *loc. cit.*, 1872, p. 15 and p. 45), larger than any of the other five, the median groove is broad; at the narrowest, at an inch from the lower end, $1\frac{1}{8}$ inch, at the middle $1\frac{3}{4}$, near the canal $1\frac{7}{8}$. It may possibly become narrower with age in Megaptera also.

The inferior ends of the condyloid cups project more than in B. musculus. This projection in the forward direction renders the cup somewhat deeper than in B. musculus; but the chief difference is the greater downward projection, by which in Megaptera the cups project below the level of the inferior arch of the bone, leaving a wide and deep notch between them. In the four mature specimens of B. musculus the rough anterior arch is seen below the level of the cups, and the notch between the slightly raised inferior ends of the cups is shallow, $\frac{1}{4}$ to $\frac{3}{4}$ inch deep, and about 3 inches wide, but in the largest atlas (the Wick specimen) considerably narrower. In the Megaptera the notch is $1\frac{1}{2}$ inch deep, and, in width, $3\frac{1}{2}$ inches below, $1\frac{1}{4}$ above. In the 50-feet-long B. musculus the distinction is much less marked, the notch 1 inch deep and wider below than in Megaptera, and the cup projects below the level of the inferior arch of the bone. But allowing for the immaturity of both, as compared with the four first-mentioned specimens of B. musculus, it appears that the lower ends of the cups project considerably more in Megaptera than in B. musculus, giving both a deeper cavity and a greater projection downwards.

Posterior Aspect of the Atlas.—The chief characters on this aspect of the atlas are (*a*) the presence of a mesial articular surface, dividing into three parts what forms one great horse-shoe articular surface in other finners, and (*b*) the form of the ligamentous area.

(*a*) *Mesial Articular Surface.*—The position and characters of this surface are seen in fig. 17. It occupies the whole height and breadth of the inferior arch of the bone; in form resembling the upper half of a blunt oval, height $2\frac{1}{2}$ inches, width 3 inches, vertically convex on its upper half, on its lower half a little concave. It is bounded below by the triangular subaxial peak, here ½ inch in length, which meets it at nearly a right angle. This aspect of the subaxial peak has somewhat the appearance of having on it an articular facet the size of the end of a finger, but there is no corresponding facet on the axis.

The mesial articular surface is separated from the great semi-lunar lateral surface by a furrow ¼ to ⅜ inch in breadth; in length 1½ inch on the right side, ¾ inch on the left. Although the separation of the mesial surface is very marked on the macerated bone, it does not follow that it was separated by ligament, or even that the articular cartilage was not continuous with that on the lateral surfaces. The floor of the furrow is not rough, and the levels of the surfaces on either side are nearly the same, the surface on the outer side of the furrow a little more projecting.[1]

(*b*) The *ligamentous area*, for the attachment of the great interosseous ligament on each side between the atlas and axis, crescentic in form in all the specimens of B. musculus, is quadrate or rhomboidal in Megaptera; also much broader, and altogether considerably larger than in B. musculus (see fig. 17, and, for comparison, fig. 5, Plate II., *loc. cit.*, 1872, showing the form in B. musculus, and also the transverse ligament). In Megaptera the length of the area averages 3½ inches, breadth about 2¼, at its lower part 2 inches. The greater breadth and quadrate form in Megaptera are gained above by its throwing out a superior external angle, prolonged like the point of a finger; below, by its extending as an inferior external angle between the mesial and lateral articular surfaces; and internally by the boundary of the canal being less curved outwards than in B.

[1] Professor Flower has described (*Proc. Zool. Soc.*, 1864, p. 402), on the atlas of the fin-whale in the Leyden Museum, taken on the north-west coast of Java, the lateral articular surfaces as not confluent below, but having between them "a distinct, oval, transversely elongated facet, and another smaller round one is situated on the upper surface of a pointed triangular projection from the hinder border of the inferior surface of the bone, which runs under the body of the axis." Also corresponding surfaces on the axis. The interval, shown in his figure (fig. 12) of the axis, between the median facet and the great lateral surface is consider-able. In my B. borealis the two lateral surfaces are confluent, as in B. musculus, and there is no articular facet on the subaxial peak. In my figure of the posterior surface of the atlas of B. musculus (*loc. cit.*, 1872, fig. v.) a narrow median furrow is seen separating the great lateral articular surfaces, but reasons are given (*loc. cit.*, pp. 15 and 50) for not regarding that as implying non-continuity of the articular cartilage.

musculus. All round its upper, outer, and inferior margins the area is defined by the sharp edges of the articular surfaces. The sharp edge crosses the upper end of the furrow between the mesial and lateral articular surfaces, and the surface of the furrow resembles that of the articular surfaces, not that of the ligamentous area. The surface of the area is undulating and rough, excavated along the outer half, deeply pitted at the superior external angle, especially on the left side, less irregular but more rough along its inner half. Internally it is bounded by a sharp edge where it meets, at a right angle, the narrow surface bounding the canal. In the specimens of B. musculus this angle is rounded off.

The *great lateral articular surfaces* present less general convexity than in any of my five specimens of B. musculus. Internally the surface is encroached on by the ligamentous area, and on the outer half it is either nearly flat or a little concave transversely. In the specimens of B. musculus the surface has a marked transverse convexity, with a little concavity towards the outer part near the raised outer edge. Instead of the more or less raised articular surface, bounded by a sharp edge, seen in B. musculus, the upper half of the articular surface in Megaptera is, as it were, scooped out to the level of the concave posterior surface of the transverse process. The less convexity of the surface, together with the greater ligamentous area, in Megaptera would seem to indicate adaptation to less movement between the atlas and axis and more firm binding of these two vertebræ together in it than in B. musculus.

Canal of the Atlas.—The canal presents some differences from that of B. musculus. The height (7¼ inches) is less than in any of the five specimens of B. musculus (7¾ to 8 inches); in the 50-feet-long B. musculus almost 8. This is probably owing to the height of the inferior arch (3⅔ inches on the anterior aspect), which is about ¾ inch greater in Megaptera than in the 50-feet-long B. musculus. The lower end of the canal has thus a less pointed form than in B. musculus. The lower, or odontoid, part of the canal, marked off from the neural part by the constriction, varies a good deal in breadth in the different specimens of B. musculus. Especially in two of the specimens previously noted (*loc. cit.*, 1872) this part of the canal is much narrower than in Megaptera, but in the 50-feet-long B. musculus it is the reverse; breadth at the narrowest part—in Megaptera 3 inches, in the B. musculus $3\frac{1}{2}$; at the widest part below—in Megaptera $3\frac{1}{4}$, in the B. musculus $4\frac{1}{8}$. Between these points is the place of attachment of the transverse ligament. It may be that the narrowing of this part of the canal depends on age.

Parts on the Neural Arch of the Atlas.—The *true articular processes* which I described (1872, *loc. cit.*, p. 39) in B. musculus are present in this 50-feet-long B. musculus, oval facets $1\frac{3}{4}$ inch transversely by $1\frac{1}{4}$ longitudinally; those of the atlas received obliquely between those of the axis, the typical relation of true zygomal processes. In Megaptera there is no trace of such articular contact, or of processes on the atlas, but distinct and large processes project here from the *axis* in the same position as the projections on which the articular facets occur in B. musculus. They are about $1\frac{1}{4}$ inch in length, and fully 3 inches broad at the base, narrower and rounded off at the end. When the bones are placed in position, these anterior articular processes of the axis overlap the lamina of the atlas for fully 1 inch, but there is no actual contact, a space of about $\frac{1}{4}$ inch intervening between them and the arch of the atlas.

Transverse Foramen of the Atlas.—In all the specimens of B. musculus the outer opening of this foramen or canal is oval vertically, the lower and outer end prolonged as a groove. In Megaptera the oval is nearly reversed, the ends anterior and posterior. This is owing to the roof of the canal being continued farther outwards in Megaptera. From the same cause the canal is longer by half an inch in Megaptera ($1\frac{3}{4}$ inch) than in the 50-feet-long B. musculus. It is also rather smaller, in Megaptera admitting the forefinger, in B. musculus admitting the thumb. In all the specimens of B. musculus the bridge completing the canal is arched forwards, where it joins the posterior end of the condyloid cup, while in Megaptera the anterior margin of the lamina is almost straight from the end of the articular cup to the spine. The arching forwards is not owing to greater thickness of the bridge in B. musculus, the thickness being about the same ($\frac{3}{4}$ inch) in Megaptera as in the 50-feet-long B. musculus.

Spinous Process of the Atlas.—There are marked differences in the spine. In all the specimens of B. musculus, the spine, besides being low, is much more developed on the posterior half of the arch, in two of them very little developed on the anterior half. In Megaptera it is higher ($1\frac{1}{4}$ inch, against about $\frac{1}{4}$ inch in the B. musculus), is semicircular in form, and rather better developed on the anterior than on the posterior half. In connection with this more anterior development of the spine is to be noted the straightness of the anterior border of the lamina in Megaptera, and also its thickness ($\frac{2}{3}$ to $\frac{5}{8}$ inch), while in B. musculus it is a depressed sharp border.

Transverse Processes of the Atlas.—While the transverse processes of the specimens of B. musculus differ from each other in detail, those of Megaptera have characters distinct from them all. The difference is mainly in their shortness (in Megaptera $3\frac{3}{4}$ inches, in the 50-feet-long B. musculus $6\frac{1}{4}$). The farther out position of the internal intertransverse tubercle (external to the line of the outer edge of the condyloid cup, and at about the middle of the upper edge of the process) gives the process a more square shape in Megaptera, as seen antero-posteriorly. The process is also broader externally in Megaptera, but this is mainly owing to its wanting the outer half, which forms the tapering, bent-back, and twisted part of the process in B. musculus. The breadth (height) of the process at the middle, at the internal intertransverse tubercle, is $4\frac{1}{2}$ inches, being greater than the length. Another character, also contributing to the square appearance of the process, is the abruptness of the junction of its lower border with the side of the body, compared with the gradual sloping-down of this junction in B. musculus. It should be added that the transverse processes of this Megaptera have been covered with cartilage, externally and half-way along the upper border, while those of the B. musculus have very little of the appearance of incomplete ossification.

32. THE AXIS.—The *anterior aspect of the body* presents articular surfaces and ligamentous markings corresponding to those above described on the posterior surface of the atlas. The ligamentous area is not sharply marked off from the low odontoid elevation, but is discernible on comparing it with that area on the atlas. The mesial articular surface is fully as well marked off as on the atlas. It curves up to the top of the

front of the broad low odontoid, with rather a narrower ending than the corresponding surface on the atlas. The furrow between it and the right lateral articular surface resembles very much the corresponding one on the atlas, in its breadth and in the character of its surface; that on the left side is very shallow, ⅛ inch in breadth at the middle, widening upwards and downwards, but well defined along its margins, and its surface resembles that of the ligamentous area rather than bone that had been covered by cartilage.

The breadth of the entire odontoid and ligamentous area is 7 inches in Megaptera, 5½ in the B. musculus. The breadth of the entire upper surface of the body is ¼ inch less in Megaptera (13¼) than in the B. musculus. The less depth of the lateral articular surfaces in Megaptera is marked.

Transverse Processes of the Axis.—The axis of Megaptera is easily distinguished from that of B. musculus by the transverse processes. In B. musculus the upper and lower processes unite to complete the ovoid ring and form a great common terminal plate external to the ring. In this Megaptera the ends of the upper and lower processes are 3 inches apart. The ends have been covered by cartilage.[1] They incline a little to each other, mainly by curvature of the upper process. Were the ring to be completed by the up-growing of the inferior process, the ring would have very nearly the same diameters as in the B. musculus.

But, even short as they are, the transverse processes of Megaptera differ materially from those of B. musculus in form and direction. The *inferior process* has the following dimensions; in B. musculus, at the narrowest part, height 3⅜ inches, thickness 1¾; in Megaptera the corresponding measurements are 2 inches and 2¼. The *superior process*, on the contrary, in Megaptera exceeds that of B. musculus both in height and in thickness. Viewed antero-posteriorly, the superior process in B. musculus is only about half the breadth (height) of the

---

[1] The soft tissue completing the ring of the axis of a fœtal Megaptera was found by Eschricht (*loc. cit.*, p. 133) to be cartilaginous. In the 46-feet-long skeleton of Megaptera in the Brussels Museum, noticed by Professor Flower (*Proc. Zool. Soc.*, 1864, p. 416) as "a nearly adult individual," the transverse processes of the axis are noted as "short, thick and convergent, but still with a wide space between them."

inferior, while in Megaptera the superior process is rather broader than the inferior.

Viewed from the side, the direction of both processes in Megaptera is seen to be downwards, and the axis of the incomplete foramen is seen to have that direction in a marked degree. In B. musculus the direction of the processes is nearly transverse, and that of the axis of the ovoid ring outwards and a little upwards. They differ also in the amount of the backward slope. The surfaces of the upper and lower processes in B. musculus are nearly on the same vertical plane, but in Megaptera the lower process is on a plane anterior to the upper process, partly at its junction with the body, and wholly so externally, from its sloping much less backwards than the upper process does.

The adaptations of these differences appear to be that, in B. musculus, the great breadth and flatness of the inferior process, and the two processes being on the same plane, are related to their having to support a great broad wing; and that, in Megaptera, the more backward slope of the upper process, as compared with the lower, is in correspondence with the slope of the same processes of the vertebræ behind it. The exception is rather in the less size and rounded form, as well as the less backward direction of the lower process in Megaptera. Its circumference is 1 inch less than that of B. musculus; the circumference of the upper process is 1 inch greater in Megaptera than in B. musculus. In its rounded form and free termination the lower process in Megaptera resembles the corresponding part of that process in Mysticetus much more than in B. musculus.

Spinous Process of the Axis.—The region of the spine of the axis in Megaptera is very different from that of B. musculus. In the latter there is the great square-shaped mass, formed by the two thick lateral longitudinal ridges, and in the valley between them the low median ridge, the true spinous process, scarcely rising to the level of the lateral ridges. The lateral ridges proceed backwards from the projections on which the true anterior articular processes are situated. In Megaptera the lateral ridges are low, proceeding backwards from the anterior quasi-articular processes noted above with the atlas.

The arch rises to the base of the spine, which projects about
1 inch as a median crest, extending the whole length of the
arch. It has a curved form, not unlike that of the spine of the
atlas, but the greater prominence on the anterior half is more
marked than on the atlas. The top of the spine is 1¼ inch
above the level of the lateral ridges.

Posterior Aspect of the Body of the Axis.—The greater
diminution on the posterior surface of the body of the axis in
Megaptera is remarkable. On the anterior aspect the total
width of the body, to the outer edge of the articular surfaces, was
noted above as only ¼ inch less in Megaptera (13¼ inches) than
in the B. musculus. On the posterior aspect, the width of the
body is in the B. musculus 11½ inches, in Megaptera only 9½.
This is a change to the lesser width of the bodies of the five
posterior cervical vertebræ in Megaptera compared with B.
musculus.

33. THE FIVE POSTERIOR CERVICAL VERTEBRÆ.—*Bodies.*—
The measurements are given in Table II. The forward diminution in *breadth* (already noted with the bodies of the lumbar
and dorsal vertebræ) ceases with the 7th cervical. The breadth
is then the same (8½ inches) on to the 3rd cervical, where it
becomes 9¼. The forward diminution in *height* is arrested from
the 1st dorsal to the 5th cervical, on which the height exceeds
the 7 inches, and diminution is resumed on the 4th and 3rd
cervical. The forward diminution in *length* goes on steadily
along the neck, from 2¾ inches on the 1st dorsal to 1½ inch
on the 3rd cervical.

[In *B. musculus* (Table III.) the forward diminution in *breadth*
ceases on the 7th dorsal. From 9⅞ inches there the breadth has
increased at the 7th cervical to 11½, and at the 3rd cervical to 11⅝
inches. The forward diminution in *height* ceases at the 5th dorsal;
from there to the 7th and 6th cervical the height has increased from
7 inches to 7⅞, and thence to the 3rd cervical has again diminished
to 7 inches. The forward diminution in *length* is continued in the
neck, from 3½ inches on the 1st dorsal to 2⅜ on the 7th cervical, and
to 1¾ on the 4th and 3rd.]

The contrast between the bodies of the cervical vertebræ in
Megaptera and B. musculus is, in Megaptera, their less breadth
compared with the height. Taking the 4th cervical, these
measurements are, in Megaptera, breadth 8½ inches, height 7;

in B. musculus, breadth $11\frac{5}{8}$, height $7\frac{3}{8}$. At the 13th dorsal
the breadth exceeds the height in Megaptera by $1\frac{7}{8}$ inch, in B.
musculus by 3 inches. At the 7th dorsal the excess is, in
Megaptera 2 inches, in B. musculus $2\frac{5}{8}$. At the 4th cervical the
excess is, in Megaptera $1\frac{1}{2}$ inch, in B. musculus $4\frac{1}{4}$ inches.
It would best express the difference to say that it is in the increased breadth of the bodies in B. musculus. In the neck of
Megaptera the bodies remain nearly the same in breadth and in
height as at the anterior dorsal, while in B. musculus, from the
7th dorsal forwards the breadth goes on increasing, the height
but little so. This increase of the breadth of the cervical bodies
in B. musculus may be related to the greater development of
tranverse processes in it than in Megaptera.

The less actual *length* of the bodies in Megaptera is seen in
Tables II. and III. In Table I. the length of the neck is given
as the same in Megaptera and B. musculus (19 inches). This
was taken as the vertebral columns lay together. The cervical
vertebræ, as now built up and standing together, have the length
of 17 inches in Megaptera, in B. musculus $17\frac{3}{4}$. The five
posterior vertebræ together are $9\frac{1}{2}$ inches in Megaptera, in B.
musculus 11. This is without the fibro-cartilages. The five
posterior vertebræ, therefore, contribute less proportionally to
the length of the neck in Megaptera than in B. musculus, and
may be regarded, so far, as in a somewhat more reduced condition. Regarded, however, in relation to the different total
length (40 feet and 50 feet) of the two carcases, not only is the
whole neck proportionally longer in Megaptera than in B.
musculus, but even the five posterior vertebræ contribute to the
greater proportionate length of the neck.

Inferior Transverse Processes.—These are present only on
the 3rd and 4th. The angular eminence on the 5th, where the
front and side of the body meet, can scarcely be termed a process. At the root they occupy nearly the whole length of
the body, not nearer the front than the back, and have very
little vertical expansion where they join the body. In these
respects they contrast with the roots of the processes in B.
musculus. That of the *third* vertebra, $3\frac{1}{2}$ inches in length,
shows on its outer half the commencement of the tubercular
stage by moderate expansion of the lower part of the process.

That of the *fourth* vertebra is shorter than the third by ¾ inch, and is more robust throughout, especially vertically, giving it a more rounded form. The tubercular stage is seen on its outer third. Viewed from below, the inferior processes stand out transversely. Viewed from the side, they are seen to be directed considerably more downwards than are the corresponding parts of the same vertebræ of B. musculus. The shortness of the inferior transverse processes in Megaptera, and their cessation after the 4th vertebra, indicate a much less development of the inferior intertransverse ligament in Megaptera than in B. musculus.

[In this 50-feet-long *B. musculus* the inferior processes of the 3rd, 4th, and 5th vertebræ show the root stage, tubercular stage, and nerve-groove stage (described *loc. cit.*, 1872, pp. 6 and 25), and the as yet incomplete terminal plate. The process of the 5th is, as in the other specimens, the strongest. The processes of the 6th show the tubercular stage, 2½ to 3 inches in length, beginning by a forward angular projection, and tapering outwards to a blunt point. The 7th vertebra shows only the low tubercle on the posterior half of the body.]

Superior Transverse Processes.—These differ from the transverse processes of B. musculus in commencement, length, direction, form, and in the absence of marked division into stages. The *commencement* of the process in Megaptera is more external on its upper margin than on its lower; in B. musculus it is the reverse to a marked extent. This is owing to the more outward position of the articular processes in Megaptera. On their lower margin the processes begin external to the plane of the side of the bodies (½ to 1 inch, increasing backwards), owing to the narrowness of the bodies and the greater length and obliquity of the pedicle than in B. musculus. In the latter the lower margin of the process begins internal to the plane of the side of the body on the third and 4th vertebræ, but not when the 6th and 7th are reached.

In *length* these five processes differ but little from each other (see Table V.). They have all had cartilage on the end, and terminate in blunt rounded ends. In *direction* their comparative shortness lessens the appearance of great convergence presented by B. musculus. Taking the distance between the processes from the 7th to the axis, at their roots and at

their tips, the convergence in B. musculus is from 9 inches at
the roots to 2¼ inches at the tips; in Megaptera from 7 inches
at the roots to 4 inches at the tips. The ends are not in contact, separated by intervals of from ¼ to ½ inch, but are nearer
each other than are the processes of B. musculus at the same
distance from the bodies. The process of the 5th, as in B.
musculus, is the central one to which the others converge, but
the 6th and 7th have not so much forward slope as in B.
musculus.

In *form* the distinction between the nerve-groove stage and
tubercular stage is scarcely recognisable. The whole process is
more rounded than in B. musculus, standing out like a long
finger, but is still somewhat flattened, surfaces forwards and
backwards, especially towards the root. The greater breadth
and flatness of the superior as well as of the inferior processes
in B. musculus may be regarded as related to their having to
support large terminal plates. The special roughnesses seen
in B. musculus for the attachment of the superior intertransverse ligament, are but faintly marked in Megaptera. All this
points to less necessity for binding together of the vertebræ of
the neck in Megaptera. Although these superior processes
might bear a ligament of considerable strength, Megaptera
wants the binding of the enormously strong external intertransverse ligaments (*loc. cit.*, 1872, p. 9) which hold together,
at the apex of the pyramid, the expanded terminal plates, and
tie them to the vast expansion of the wing of the axis.

Comparing the superior processes with each other, the 3rd
is the most slender in both Megaptera and B. musculus, but
more strikingly so in Megaptera. In B. musculus they increase
moderately in strength backwards from the 3rd to the 5th, and
then greatly on the 6th and 7th. In Megaptera that progres-
sion is interrupted by the 4th process being as thick as the 5th
on the left side, and much thicker than it on the right side.
The 6th has the same robustness as the right 4th. The 7th
undergoes very sudden enlargement, and, relatively to the others
and to the 1st dorsal, is larger than in B. musculus. It is
almost, if not quite, as large as the 1st dorsal, and presents
almost as large and as thick an outer end. The 7th stands out
as a strong process, and in the recess between it and the axis

are the free ends of the four intervening processes in a gradually increasing line backwards; the second 1½ inch less projecting than the axis, the 6th only a little less projecting than the 7th.

Another differential character of the superior transverse processes of Megaptera is their straightness. In B. musculus they curve upwards, so as to give the ring a decidedly concave upper boundary; in Megaptera they are almost straight from where they leave the pedicle, with a little concavity at their outer end only. The 7th begins to show a little general concavity. This straightness is seen also on the upper aspect, though with less definite outline, and contrasts with their convexity in B. musculus.

In regard to the size of the space between the upper and lower processes, it is seen in Table V. to be about the same in both, a little less in Megaptera. The distance between the ends of the two processes of the 3rd and 4th vertebræ is 5½ and 6 inches. The ends of the upper processes are on a plane posterior to those of the lower, to the following extent—axis, 2 inches; 3rd vertebra, 1½; 4th vertebra, 1 inch. In B. musculus they are on the same level, having to meet externally; but on the 6th, as the lower process stops short, the outer end of the upper process is on a plane somewhat anterior to the end of the lower process. In the fœtus of Megaptera, Eschricht found the soft tissue completing the rings behind the axis to be fibrous.

The comparison of the transverse processes in B. musculus and Megaptera would seem to show that the presence of complete processes, forming rings, is not in adaptation to the protection of the great vascular rete that occupies the space, but for the attachment of ligaments and muscles.

Articular Processes of the Five Posterior Cervical Vertebræ.—The measurements in Tables II. and III. show the much greater width apart of the articular processes in Megaptera than in B. musculus, the distance averaging about 11 to 11½ inches in Megaptera, in B. musculus about 9 to 9½, the measurements taken from the outer edge of the processes. The great increase (going forward) in width apart between the processes begins in both at the 5th dorsal, owing to the turning outwards of the processes; obtains its maximum on the 1st

dorsal in B. musculus (10 inches), in Megaptera on the 1st
dorsal and 7th cervical (11½), and thence, along the neck,
diminishes a very little forwards to the 3rd. This greater
width apart of the articular processes in Megaptera, by 2 inches,
is the more remarkable, as the bodies are narrower in it than in
B. musculus by fully 2 inches. From this it results that, when
the two sets of cervical vertebræ are viewed from below or from
above, the sides of the bodies are seen to be considerably internal to the outer edge of the articular processes in Megaptera,
and considerably external to them in B. musculus, affording a
very distinctive character.

The above characters are seen in the articulated position.
The following characters of the articular processes are seen
when the vertebræ are laid out separately alongside each other.
The articular *surfaces* are larger and ovoid in form in Megaptera, averaging 2 inches transversely by 1½ antero-posteriorly;
in B. musculus they average 2 inches transversely by ¾ to 1
inch antero-posteriorly. They are larger between the axis and
3rd (in Megaptera 2½ by 2, in B. musculus 2¼ by 1¼), and become
smaller and more irregular backwards along the series in both.
The difference in form is owing to the anterior processes projecting more in Megaptera; so that in B. musculus the anterior
half of the ovoid is wanting, especially on the inner side, where
the border of the process falls gradually into continuity with the
anterior border of the lamina.

In Megaptera the anterior processes, from the 3rd to the 6th
vertebra, are moderately convex, becoming flat on the 7th; in
B. musculus the anterior processes are convex on the 3rd, 4th,
and 5th, flat on the 6th, concave on the 7th. Their convexity
on these three vertebræ in B. musculus is owing to their bending
down towards the transverse process; while, behind the 5th,
the upper edge of the transverse process comes quite up to the
outer edge of the articular process.

Considering the firm binding together of the bodies by their
fibro-cartilages, there can be very little movement at these
diarthrodial surfaces, and most of them present irregularities
of surface. But their greater extent in Megaptera would seem
to indicate more movement at the articular surfaces in it than
in B. musculus.

Viewed *in relation to the pedicle,* a distinctive character is afforded by lines drawn vertically up from the inner and outer **borders of the pedicle.** **The inner** line, drawn from the concavity of the border, corresponds in Megaptera to the inner end of the articular process; in B. musculus it cuts off about the inner third of the process. The outer line, drawn from where **the** pedicle **leaves the body, would, in** Megaptera, cut off the inner third or fourth of the articular surface less and less backwards, so that on the 6th and the 7th line would fall at the inner end of the process. In B. musculus the line would pass at the outer edge of the process of the 3rd, and more and more external to the processes as we go back. These differences result partly from the articular process being in part placed in B. musculus on the lamina, while in Megaptera they are placed above the pedicle, and above the root **of the transverse process;** **and partly from the greater breadth of the pedicle in B. mus-**culus.

Pedicles of the Five Posterior Cervical Vertebræ.—As **seen** in Table II., the diminution in the *breadth* of the **pedicles in** Megaptera, as we go forward in the dorsal region, is continued **to the 6th** cervical ($1\frac{5}{8}$ inch), in front of which there is very little **change.** The forward diminution in the *thickness* of the pedicle, from the dorsal region, ceases after the 6th cervical ($\frac{7}{8}$ inch), except on the 3rd cervical, on which it is slightly increased. The greater size of the anterior articular process may account for the pedicle of the 3rd being a little stronger than that of the three behind it, although its transverse process is **the** weakest.

[The much **greater** *breadth* of the pedicles in *B. musculus* is **related** to the greater size of the transverse processes they have to support. Their relation to the sides of the bodies is much the same in both, and the width of the canal is nearly the same in both ; their greater **width in B.** musculus is gained by the greater breadth of the bodies in it. Their forward narrowing goes on from $3\frac{3}{4}$ inches on the 1st dorsal, and $3\frac{1}{2}$ on the 7th cervical, to $2\frac{1}{2}$ inches on the 3rd cervical. But their *thickness* is less than in Megaptera, $\frac{5}{8}$ inch against $\frac{7}{8}$.]

The appearance of less *height* of the pedicle in B. musculus is deceptive, owing to its greater breadth, and to the breadth and lowness of its connection with the transverse process. Measured along the middle to **the middle** of the anterior articular process,

L

the height is very nearly the same in both. The high-up commencement of the lower border of the transverse process gives a
longer neck to the pedicle in Megaptera, and the inward slope
of that border towards the body increases the upward and outward obliquity of the pedicle in Megaptera. That obliquity is
related to the more outward position of the articular processes
in Megaptera.

Spinal Canal in the Neck.—(*a*) *Capacity.*—The increase in
height, with diminution in width of the canal at the 3rd vertebra, as compared with the 4th, in both Megaptera and B.
musculus, is a transition to the form in the axis, in which the
height is considerably increased and the width diminished. In
height the canal increases a little backwards from the 4th to
the 7th (3⅛ inches to 3⅔), after which the increase to the dorsal
height begins (1st dorsal, 3⅝ inches). The *width* is greatest at
the 6th and 7th cervical (7 inches) and 1st dorsal (6⅞ inches),
and thence diminishes forwards to 6½ inches on the 4th, and
backwards along the rest of the column.

[In *B. musculus*, the *height* from the 4th to the 7th cervical (2⅞
inches) is less than in Megaptera (1st dorsal, 3 inches). The *width*,
from the 4th to the 7th cervical, increases from 6⅝ inches to 7; is 7 1/16
on the 1st dorsal, and thence diminishes backwards.[1]]

(*b*) *Form.*—The somewhat higher arch in Megaptera is not
obtained by greater slope of the lamina towards the spine, but
rather by the lateral angles being carried higher up, and therefore less sharp than in B. musculus. In the floor of the canal
the longitudinal median ridge of the bodies is seen *in B. musculus*, extending all along the neck. It begins near the fore

---

[1] The most capacious part of the spinal canal being in both the back part of
the neck, would accord with an assumed enlargement of the spinal cord at the
origin of the nerves of the pectoral fin. If so, the enlargement should be greater
in Megaptera with its enormous pectoral fin. The bony canal here is more
capacious in Megaptera, but only from its greater height. In the measurements
of the three previously noted series of cervical vertebræ of B. musculus (*loc. cit.*,
1872, Table, p. 20) the capacity of the canal of the 6th and 7th was found not to
be greater than at the 4th and 5th cervical. In one it was greatest at the 5th
and 6th; in a second at the 4th and 5th; in the third, at the 6th and 7th. But
considering the small size of the spinal cord (at the middle of the neck the tube
of dura mater was found to be only 1 inch in diameter (*loc. cit.*, 1872, p. 5), and
that the great part of the bony canal is occupied by rete mirabile, correspondence
between any special enlargement of the cord and the capacity of the bony canal
can hardly be expected.

part of the axis, is 1 inch broad on the following more anterior vertebræ, and becomes less marked on the more posterior. The body is concave transversely on each side of it, so that, but for the ridge, the whole upper surface of all the bodies would be transversely concave. In *Megaptera* the median ridge is less marked. It begins at the same place on the axis, is narrow on the 3rd, mesially grooved on the 4th, less marked on the 5th, and very faint on the 7th. Transverse concavity on the upper margins of the bodies ceases on the 5th, and on the 7th the whole upper surface of the body is convex transversely (continued on the dorsal), presenting a marked contrast to the 7th of this B. musculus.

Spinous Processes of the Five Posterior Cervical Vertebræ.— The 3rd and 6th have been injured. The 4th ($\frac{2}{3}$ inch in length) and the 5th ($\frac{3}{4}$ inch) are a little longer, the 7th ($1\frac{1}{2}$ inch) a little shorter than in B. musculus. The 7th has a backward direction in Megaptera, like the anterior dorsal spines; in B. musculus it is nearly straight up, like that of the 1st dorsal. After the 4th, the spines in Megaptera have a more unfinished appearance than those of the B. musculus.

Laminæ.—The laminæ are necessarily longer (transversely) in Megaptera, from the farther-out position of the articular processes, but the chief differential characters are their breadth (antero-posteriorly), filling up the spaces, and their much greater thickness and convexity on the upper surface, in Megaptera. The convexity appears as a rough bulge occupying about the middle half of the lamina transversely, and the posterior $\frac{3}{4}$ of the breadth. The same is seen on the 1st dorsal and less on the 2nd. These prominences seem to correspond to the anapophysial processes previously described in B. musculus (*loc cit.,* p. 24, and fig. 3, 1872), but they are not prolonged behind the lamina. They are rough on the surface and evidently for the attachment of muscular or ligamentous structures.

The laminæ of the 3rd and 6th vertebræ have been injured from the rough usage they received in Dundee harbour, but, as far as can be judged now, the spaces have been filled up by the anterior overlapping the posterior, except on either side of the 3rd arch, although it cannot be determined how far this is from breakage of the thin anterior edge of the 3rd and 4th lamina.

On separating the vertebræ, a fossa is seen on the posterior aspect of each thickened part of the laminæ, as if the anterior edge of the lamina behind had passed into a cavity on the lamina in front. On the 5th, on which these fossæ are most pronounced, they are elliptical, 2 inches transversely by $\frac{1}{4}$ to $\frac{1}{2}$ inch in breadth, the broader end outwards, at about 1 inch internal to the articular process, the narrow end 1 inch or less from the spine; depth, from $\frac{1}{6}$ to $\frac{1}{8}$ inch or less. They have a very distinct and raised inferior margin, which is the true lower edge of the under surface of the lamina, the fossa itself being on the thickened posterior margin of the lamina. They have not the appearance of having been covered by cartilage, but are evidently parts into which something definite has been received. On the 3rd, it is seen, on the less injured side, as a shallow depression near the spine, 1 inch in length. On the 4th, they have the length and breadth noted above on the 5th, but are divided by a median ridge into an inner and outer part, as if two parts of the lamina behind had been lodged in them. On the 6th, it is seen on the least injured side, as if divided into three parts. If the corresponding part is represented on the 7th vertebra, it is as a rough depression running along the posterior margin of the lamina, beginning at the articular process, seen on the upper aspect of the posterior margin of the lamina. So far as they are complete, there is no special thickening, or mark, on the anterior edge of the lamina behind corresponding to these fossæ.

[In the *B. musculus* there are spaces on each side of the 7th and 6th arches, wide enough to receive the hand flat, and a narrower one between the 5th and 4th. The laminæ are flat and thin compared with those of Megaptera, the anterior border of the 3rd, 4th, and 5th so thin as to be flexible. The laminæ are not nearly so thin as in the 65 to 66-feet-long B. musculus, but are thinner than in the 64-feet-long B. musculus before described (*loc. cit.*, 1872). Nor are the very marked anapophysial processes described in them present in this 50-feet-long B. musculus to any great extent. They are seen only on the 3rd and 4th vertebræ, with a mere trace on the 5th, and are short and flat. The fossæ seen on the posterior border of the laminæ in Megaptera are not present in this 50-feet-long B. musculus, but corresponding fossæ or grooves are seen, more or less, in the three more mature individuals, best marked in the 64-feet-long one, least marked in the 60$\frac{1}{2}$-feet-long one. They are not so sharply marked on their inferior border as in Megaptera, and lie more towards the spinous process.

Van Beneden and Gervais remark (*Ostéographie des Cétacés*, p. 132), "C'est dans la région cervicale surtout que se trouvent les différences qui séparent le keporkak de la Megaptera Lalandii." The differences which they go on to note between these two supposed different species, Megaptera longimana and Lalandii, are, so far as I can judge, not greater than those seen in the four series of the cervical vertebræ of B. musculus above referred to.]

THE RIBS.

The following table (Table VI.) shows the proportions of the ribs in comparison with those of the 50-feet-long B. musculus:—

34. TABLE VI.—*Measurements of the Ribs of Megaptera longimana and of the 50-feet-long B. musculus, given in inches.*

	Megaptera longimana.						B. musculus.					
	Length.[1]	Depth of Curve.	Breadth at Middle.	Thickness at Middle.	Angle to Tubercle.[2]	Tubercle to end.[3]	Length.[1]	Depth of Curve.	Breadth at Middle.	Thickness at Middle.	Angle to Tubercle.[2]	Tubercle to end.[3]
1st Rib, . .	41	8	4½	1⅞	3	3	36	9	4¼	1¼	3¼	1½
2nd „ . .	53½	13½	3½	1⅞	6	1¾	50½	12¼	4	1⅝	3¾	6
3rd „ . .	59	16½	3	1⅕	4¾	2¼	60	14½	3¾	1½	5	5¾
4th „ . .	61	17¼	2⅞	2	5	2¼	63½	16½	3	1½	4½	4
5th „ . .	62	19	2½	1⅞	6	2	66¾	17¾	2¾	1½	5	2½
6th „ . .	63*	19 1/16	2¾	2	5¼	2	67	18¾	2⅝	1½	5	2½
7th „ . .	59½	18½	2½	2	5½	2	66	17½	2½	1¾	5	2
8th „ . .	56½	18	2⅝	2	...	2¼	69¾	16¾	2¼	1½	5	1¾
9th „ . .	55	16⅝	2⅝	2	...	2¼	61	14½	2½	1½	4½	1½
10th „ . .	53¾	14¾	2¼	2	...	2⅜	58	13¼	2 1/16	1 3/16	4½	1½
11th „ . .	51	11½	2	1⅝	...	2¼	54½	11	1½	1½	4½	1½
12th „ . .	49⅝	10	2½	1¼	...	2¼	51½	8¼	1½	1½	3	1½
13th „ . .	49¼	8¼	2	1½	...	1¼	51	5½	2	1	...	1¼
14th „ . .	45¼	7½	1½	1½	...	¾	51¼	5¼	2¼	¼	...	...
15th „ . .	...	...	...	...	...	...	32½	3	1¼	¾	...	...

35. GENERAL AND DIFFERENTIAL CHARACTERS.—*Length.*—All the ribs of the Megaptera are shorter than those of the B. musculus except the two first. The greater length of especially the first in Megaptera is considerable. The lengths taken along the outer border are, of the first rib, Megaptera 48 inches, B. musculus 44¼; of the second rib, Megaptera 65 inches, B. musculus 60. The longest rib in both is the 6th.

[1] From top of tubercle to farthest part of lower end.

[2] From top of angle to top of tubercle.

[3] Transversely, from innermost point to opposite the top of the tubercle.

* The weight of the 6th rib is in Megaptera 228 ounces, in the B. musculus 148 ounces.

Taken along the outer border **the lengths of the 6th** ribs are, Megaptera 80 inches, B. musculus 82.

Breadth.—The greater breadth of the *first rib* in Megaptera is on the right side only and not throughout, as the following measurements show, in inches :—

	Megaptera.		B. musculus.	
	Right.	Left.	Right.	Left.
At external neck,	3	3½	4½	same
At middle of upper half,	3½	3½	3¾	same
At middle,	4½	3½	4¼	same
At middle of lower half,	6	4½	5¼	5¾
At 6 inches from lower end, top of sternal notch,	7½	5	6	6½
At lower end,	3½	2¾	7¾	8½

On its lowest 6 inches the first rib is deeply excavated anteriorly on more than half of its breadth (see **fig. 18**). This peculiarity belongs to its relation to the sternum. In B. musculus this notch is not present, the rib passing of full and increasing breadth in below the wing of the sternum. A shallow excavation, ½ to ¾ inch deep, is seen on the **lower part of the anterior border** of the 2nd rib in Megaptera.

On the 2nd, 3rd, and 4th the breadth is greatest in **B. musculus.** From the 7th back to the 12th, greatest in Megaptera. On the 14th Megaptera loses, being the last rib. To the eye it appears evident **that** in Megaptera the more **anterior ribs are narrower, and the** posterior ribs broader, **than in B. musculus.**

In *Megaptera* the external neck is broader than in B. musculus, except on the 1st and 2nd, and is the broadest part of the bone except on the 1st and last, on the 2nd and 3rd very much the broadest part; all the shafts are narrowest at their lower end except the last; the broadest part of the shafts is below the middle, on the third quarter of the bone. In *B. musculus* the broadest part of the shaft is at about the middle of the lower half; on the lower quarter of the bone the breadth diminishes downwards, except on the 5th and 6th ribs, which rather increase in breadth to the end; from the 4th to the 13th the breadth is greater at the middle of the upper half than at the middle of the bone. The ribs of Megaptera and B. musculus, therefore, differ in breadth at different parts thus :—In **B. musculus,** external neck narrower (except on the two first); upper half of shaft broader than at middle; broadest part lower down, about middle of lower half, while in Megaptera it is on the third quarter of the bone; taper less towards lower end, from the 2nd to the 7th or 8th, than in Megaptera except on its first and last. The breadths of the longest rib of each, the 6th, are, in Megaptera and B. musculus respectively,—at external neck, 3⅝ and 3 inches; at middle of upper half, 2⅝ and 3; at middle, 2⅝ and 2⅜; at middle of lower half, 2⅞ and 3; at 6 inches from lower end, 2⅝ and 3¼.

Thickness.—The ribs of Megaptera and B. musculus differ mainly in their greater thickness in the former, becoming marked after the 2nd and continuing to the last, as seen in the 4th column of the Table. To the grasp, they are ovoids in Megaptera, becoming, relatively to the breadth, thicker as we go back, the four last ribs nearly as thick as they are broad. The measurements given in the Table are taken at the exact middle of the entire rib, which is about the thickest part, but there is not much diminution in thickness till upon the upper and lower fourths of the bone. The upper thinning becomes marked as we approach the angle, and is greater than that towards the lower end. From the 3rd or 4th to the 9th or 10th, and again on the last, the thickness to the very end below in Megaptera is very striking compared with the corresponding ends in B. musculus.

[In *B. musculus* the beam of the 2nd rib is external, the inner ⅔ thin and deeply grooved on both surfaces; this along about the middle half of the length of the shaft. The 3rd rib the same, but the grooving less marked. On and after the 4th, the thickest part is at, or a little internal to, the middle of the breadth, with about equal sloping on each side of it. The 13th becomes thin externally, the 14th and 15th are thin across their whole breadth. On the typical ribs the beam projects chiefly on the external surface, beginning at the angle, passing obliquely to about the middle of the surface, and ceasing near the lower end. The internal part of the upper half of the shaft is strengthened by a minor beam seen on the inner surface, beginning at the inner border and reaching obliquely outwards to the main beam. The inner surface, taken generally, is flatter than the outer surface, especially along the lower half. At the posterior border a special *sub-costal groove* and ridge are present towards the lower end, from the 5th to the 12th, beginning 10 to 12 inches from the lower end of the rib and extending upwards for about the same distance; best marked on the **9th, 10th, and** 11th, where it is as broad as the little finger.]

In *Megaptera* the course of the beam is the same as in B. musculus on both surfaces, but is rendered much less distinct by the greater general thickness of the bone on each side of it. On the 2nd and 3rd ribs the beam has already reached to about the middle, and there is scarcely any of the grooving **to its inner** side which **is** so strongly marked in B. musculus on the 2nd and 3rd, and even from the 4th to the 7th above their middle. **A trace of the sub-costal groove is** seen from the **3rd** to the 10th, from 6 to 12 inches in length, a little above the junction of the lower and middle thirds of the bone, somewhat higher up on the 10th. Where it is sharp-edged it has the breadth of a goose-quill.

Curvatures of the Ribs.—Although these ribs are more simple than in most mammals, the four curvatures are seen and differ in Megaptera and B. musculus—(1) The curvature of the axis, enclosing the chest, is seen in the second column of the table. Except on the 1st rib, in which it is 1 inch less ($1\frac{1}{2}$ on the left side), the depth of the curve is seen to be greater in Megaptera than in B. musculus along the whole series, thus **giving** Megaptera a wider thoracic cavity. It increases to the **longest** rib in both. **The** greater curvature in Megaptera is most striking on its last two ribs.

Of the minor curvatures in (2) that of the *borders*, giving the sigmoid form, **the** differences **are not** very marked. **On** the outer border the concavity below the angle is seen in B. musculus from the 1st to the 11th, owing to the greater prominence of the angle; in Megaptera on the **2nd** and 3rd, and again on the last four, but not much. On the **lower third**, the concavity on **the** outer border, with convexity on **the** inner, is seen in Megaptera from the 2nd to the 6th; in B. musculus only on the last four; **on** the 13th, exceptionally well, the shaft of this rib presenting **a** decidedly sigmoid **form**. These curvatures are greatly exaggerated **on** the last **rib** of the Megaptera, and usually on the **last rib** of B. musculus, as noted below.

(3) *Curvature on the Surfaces.*—This is much **more** marked in Megaptera, both above **and below.** In Megaptera the upper

part of the rib is much bent forwards at the angle, while the whole shaft is bent with the concavity backwards, giving a marked sigmoid form when the rib is seen edgeways. In B. musculus there is, above, only a gentle bend forwards on the upper $\frac{1}{4}$, without any rapid bend at the angle; and below, a well-marked bend backwards on the lower $\frac{1}{4}$ of the shaft, giving the sigmoid form. When the 6th ribs, the longest, of Megaptera and B. musculus are laid together on the floor, resting on their outer border, the upper end rises, in Megaptera 10 inches, in B. musculus 6 inches; and the lower half of the shaft forms an arch $2\frac{1}{2}$ inches deep in Megaptera, $1\frac{1}{2}$ deep in B. musculus. The rise at the upper end in Megaptera begins about 15 inches from the articular end, but much the greater part of it takes place at the angle, about the middle of that distance. The rise in B. musculus is by a gradual sweep along the upper $\frac{1}{4}$ of the bone, not marked at any particular place. If the external neck be now made to lie level on the floor, the whole shaft in Megaptera rises as an arch 13 inches high at the middle; that of B. musculus as an arch 5 inches high, and along its lower $\frac{3}{4}$ only, highest at $\frac{1}{4}$ from the lower end. When these two ribs are turned on their outer surface, and the external neck laid level on the floor, the lower end rises from the floor, in the Megaptera rib 37 inches, in the B. musculus rib only 9 inches. The differences are also well seen when the two ribs are made to stand together vertically. These curves of the surface are much better marked on the anterior half of the series (except the 1st) than on the posterior half.

(4) *Torsion.*—The torsion or twist of the plane is seen below on the lower half of the shaft; above, on the upper part at the angle, and specially at the articular end. The *lower torsion* is very evident in B. musculus, the inner margin on the lower half of the shaft twisted forwards, so that the surface which is anterior above now looks outwards. In Megaptera, the greater thickness of the shaft renders this torsion less obvious, but it is greater than in B. musculus. The *upper torsion* is much greater in Megaptera than in B. musculus. Along with the upper bend forwards on the surface, the plane is at the same time twisted, the inner (now lower) border back-

wards. This twist is in the opposite direction to that of the lower half of the shaft, but to a less extent. The torsions thus accompany the two bends on the surface. The effect of the upper bend and torsion is to give the upper part of the rib a more horizontal direction inwards, and to render the plane of its surfaces vertical ; the effect of the lower torsion is to make the lower half of the shaft face more outwards as it sweeps obliquely back along the thoracic wall. These differences in the bend of the surfaces and in the torsion, between Megaptera and B. musculus, would seem to be adaptations in Megaptera to a more abrupt change from the transverse direction of the upper part to the oblique direction of the shaft.

The special *torsion at the articular end* is a rapid twist of its lower part backwards, affecting also the lower part of the inner half of the external neck. The result is that the articular end, where it meets the transverse process of the vertebra, is directed downwards and backwards, decidedly so in B. musculus, moderately so in Megaptera.

The Last Rib.—This rib in Megaptera has the very undulating character usually seen in the last rib in B. musculus and other finners. This is but an exaggeration of the curvatures of the other ribs, with, if the rib is long enough, the addition of a third curvature to the sigmoid form. On the posterior border, after the slight concavity external to the articular end (there is no angle), there is the great convexity occupying about the upper $\frac{2}{5}$ of the bone ; then a wide concavity, occupying about the middle $\frac{1}{3}$ of the bone. These with corresponding curvatures on the anterior border, complete the sigmoid form (like the human clavicle, but not so much bent as it), and there is only that in the 15th rib of this B. musculus, and in the last rib of my B. borealis. But in this Megaptera a third curve is present on the lower $\frac{1}{3}$ of the bone, convexity behind. The hinder margin thus presents two convexities and one concavity, the anterior margin two concavities and a convexity. The 13th rib, with much less of the upper two bends, shows the third of these bends more typically, owing to the tapering of its posterior margin towards the end. The want of this tapering on the last rib (breadth of end $2\frac{1}{2}$ inches, of end of 13th, $1\frac{1}{2}$ inches), with a slight bend back at its broad lower end, gives the last rib, of the right side, the appearance of having a short fourth bend on its lower six inches. The left 14th rib has not this, and the middle of its three curves is less pronounced than the corresponding curve on the right.

[The 15th rib of my 64-feet-long *B. musculus* (72 inches long) shows the three curves on a great scale. There was a 16th pair of ribs in that B. musculus (this *Journal*, 1871, p. 115) loose in the flesh (right 30 inches, left 22 inches long), which the 15th, the last, of this 50-feet-long B. musculus closely resembles, except that the latter has more of the sigmoid form. The upper 9 inches taper, the upper 6 inches rounded, to a point half the size of the end of the little finger.]

36. VERTEBRAL ENDS OF THE RIBS.—The modifications of these in Megaptera will be better understood after observing them in B. musculus.

[*In B. musculus.*—The 1st rib of this B. musculus has no beak.[1] The 2nd and 3rd ribs have a long capitular process or beak. This process, and the ligament which prolongs it to the body of the vertebra in front, together represent the neck and head of the complete rib of the toothed Cetacea and of most mammals. Above the base of the beak is the well-marked tubercle by which the rib articulates with the transverse process of its vertebra, and between this and the angle is the moderate constriction which may be termed the external neck. The common error of calling the end of the ordinary or beakless ribs of the whalebone whales the "head," and the constriction external to it the "neck," was emphatically remarked on long ago by Eschricht (*loc. cit.*, p. 137). The rapid shortening of the capitular process after the 3rd gives the ribs first a sloping and then a rounded end. The 4th shows a considerable slope to a sharp point; after the 4th there is less and less slope, and the angle below is rounded off, so that the most projecting part of the end is not its

---

[1] In my 60½-feet-long B. musculus the 1st pair of ribs have as well-marked and as long a beak as the 2nd pair have (this *Journal*, 1872, p. 47). In my 64-feet-long B. musculus the beak occurred on the left side as a separate piece, articulated by cartilage to the lower part of the broad end (this *Journal*, 1871, p. 116, and fig. 4, Plate VII.). The condition on the right side could not be ascertained. In this 50-feet-long B. musculus, the end of the 1st rib is broad and rounded, the lowest part, from which a beak would have proceeded, somewhat rounded off below and thin, and without any appearance of a movable beak having existed. The cartilage-covered surface extends over the whole height and breadth of the end: height 5¾ inches; breadth above, 1¼ inch, at middle ⅔ inch, at lower part ¼ inch. Upper ⅔ convex backwards, lower ¼ concave backwards. The occurrence of a beak on the 1st rib in B. musculus appears to be a matter of ossification or of variation.

lowest part. The sixth column of the Table shows the amount of the slope from the top of the tubercle to the most projecting part of the end.

The *articular surface* or cartilage-covered area[1] begins at the top, or, it may be, a little external to the top of the tubercle. On the 1st rib it occupies the whole end. On the 2nd and 3rd it occupies the inward slope of the tubercle, or what might also be called the broad part of the beak, for 4 to $4\frac{1}{2}$ inches ; leaving a narrow beak proper, 5 inches long on the 2nd rib, 4 inches long on the 3rd rib. On the 4th rib the articular area extends over the whole sloping end, and on the ribs behind the 4th goes also a little below at the rounded-off part.

Adaptation of the Ribs to the Fossæ on the Transverse Processes.— It is not easy at first to see how the two somewhat narrow surfaces are adapted, the rib-ends being mainly vertical, the fossæ mainly antero-posterior in direction. The measurements of the end of the 7th rib are—height, $3\frac{3}{4}$ inches ; breadth, upper part 1 inch, at middle $1\frac{1}{4}$, lower part $1\frac{2}{8}$. Those of the fossa on the 7th transverse process are—antero-posteriorly 4 inches, vertically 3, depth $\frac{3}{4}$ inch. If a middle rib be so placed that the long axis of the two surfaces shall correspond, the rib will be nearly horizontal. The mode of articulation, by fibrous cushion, may render exact adaptation of surfaces less necessary than in the case of diarthrodial joints, but the adaptation becomes evident on close examination. If the 7th rib is placed naturally, the lower end carried back to about opposite the 3rd transverse process behind its own, the articular end is seen to be directed downwards and backwards at an angle of about 25°, and to fit against the anterior $\frac{1}{2}$ or $\frac{2}{3}$ of the costal fossa. That is the part on which the fossa is buttressed by the thick ridge bounding it in front, thus offering resistance to the rib, and only the anterior $\frac{1}{2}$ or $\frac{2}{3}$ of the fossa have been covered with cartilage. The cartilage has been continued down upon it from the end of the transverse process. When the 12th is reached the whole area is cartilaginous, being placed on the end of the transverse process. But the rib is longer, vertically, than the fossa. Part, about $\frac{1}{4}$ or $\frac{1}{3}$, projects above the fossa, and, corresponding to this, is the concavity on the extreme edge of the transverse process, above the anterior $\frac{1}{2}$ of the costal fossa. The rounded-off part below may project under the fossa.

Looking now to the exact form of the articular end of the rib, the terminal torsion is seen to give it the downward and backward direc-

---

[1] The parts which have been covered by cartilage are easily recognised by their roughness and perforations. The costo-transverse articulation in finners appears to be, not by regularly formed diarthrodial joints, but by fibrous cushion. Within this I have found an irregular synovial cavity (this *Journal*, 1872, p. 48), but in B. rostrata I found no synovial cavity at any of the costo-transverse articulations. The cartilage on these tubercles and ends may be only the growing cartilage of the end, but it may be regarded as also articular in function.

tion, which is increased by the lower part being prolonged backwards into a blunt point. Two parts may be recognised—the upper, sloping upwards and outwards, is seen in the end view to face obliquely backwards, to be nearly vertical, and to be separated from the chief part by a gentle concavity, reaching upon it from the posterior concave margin. The part thus marked off appears to be that which projects above the fossa. The lower and chief part faces inwards, is more decidedly convex antero-posteriorly than the upper part, and has the oblique direction downwards and backwards. This lower part of the articular surface, as seen in the end view, is pointed below on the first three ribs; from the 5th to the 8th broadest below; on the next three the terminal torsion is so much diminished that the point of the lower part is downwards rather than backwards; on the 12th, 13th, and 14th the end is rounded, but greatest vertically.]

In Megaptera.—As compared with those of B. musculus, the most striking difference at the vertical end of the ribs in Megaptera, behind the 3rd, is the lowness and rounding off of the angle. The concavity between the tubercle and angle, on the external neck, is much less, is very slight after the 7th, and after the 9th or 10th is not present. The gentle elevation corresponding to the angle in B. musculus diminishes and disappears at the same stages of the series. The slightness of the constriction is owing also to the filling up of the external neck, the upper border of which, on the anterior seven, slopes a little upwards to the low angle. This gives the upper 12 inches or so of the rib a greater bend down than in B. musculus, and a broader external neck compared with the breadth of the shaft. This direction of the upper part of the rib and the lowness of the angle may be considered as related to the upward direction of the transverse processes, in below the ends of which they are received, although that direction of the transverse processes is continued farther back on the series.

The *first* rib has a rounded top, from the low angle inwards, tapering downwards and inwards to a blunt point. The inner half (3 inches) of this may be assigned to the beak (best marked on the left rib—see fig. 18), much turned back on its last $1\frac{1}{2}$ inch; but there is no distinct tubercle, and only the blunt point, about 1 inch thick, has been covered by cartilage. The *second* rib has a prominent tubercle, the end sloping obliquely downwards and inwards, giving a broad triangular beak, $1\frac{3}{4}$

inch in length. The whole of this sloping end, 5 inches, has been covered by cartilage, as all the ends behind this have been. The *third* rib has rather more slope than the second.

The extent to which the lower part of the ends project inwards beyond the upper part at the tubercle is seen in the sixth column of the Table. It continues to be about 2 inches back to the 12th. The 2nd and 3rd ribs cannot be said to have a beak in the sense that these two ribs have in B. musculus, or even as the 4th of B. musculus has. After the 7th the slope is greater in Megaptera than in B. musculus, as seen in the Table. The remark of Van Beneden and Gervais that " la troisième surtout et la quatrième different des autres par une tête distincte " (*op. cit.*, p. 127) does not apply to this Megaptera.

Although, as seen in the end view, the articular ends of the ribs are directed less downwards and backwards than in B. musculus, the terminal torsion, affecting the inner half of the external neck and the end proper, is very marked. The twist back of the lower part of the articular end is well seen when the series of ribs are laid with the posterior surface upwards. On the first four there is both bending back of the inner part of the external neck and twisting back of its lower edge. From the 5th to the 8th there is only the decided twist; on the next two less twist; and on the last four more. The twist back of the lowest part of the articular end in some (4th to 7th on right side, 4th and 5th on left side) runs on to a projection like the end of a finger. This projection is marked on the left 5th and 7th, $\frac{1}{2}$ to $\frac{3}{4}$ inch long; on the 5th it is like the end of a thumb; on the 7th like the end of a little finger. This projecting cone is not seen in the front view of the rib, and has not been covered by cartilage. It may be regarded as a rudimentary beak, but it is irregular and not symmetrical.

Adaptation of the Ribs to the Fossæ on the Transverse Processes.—The distinction of the articular end of a typical rib into upper and lower parts is more marked in Megaptera than in B. musculus. The whole area is more bent, almost kidney-shaped. The concavity, going on the surface from the most concave part of the posterior margin, marks off about the upper $\frac{1}{3}$. This part is curved backwards, is much rounded off as it ascends, and faces obliquely backwards, while the lower part ($\frac{2}{3}$) faces inwards, and is more convex in both directions. When the rib is applied naturally to the vertebræ, the concavity on the articular end corresponds to the fore part of the upper boundary of the fossa ; the upper part rises above the fossa, curving towards the back part of the upper edge of the fossa, but, as the outer edge of the transverse process is thick (1 inch at the middle, more in front, less behind) the rib does not seem as if it had risen above the level of the cartilage with which the process has been tipped ; and the lower $\frac{2}{3}$ of the rib, more ball-like, occupies the anterior part of the fossa, the axis corresponding to that of the fossa, the direction downwards and backwards. The measurements of the end of the 7th rib are—height, almost 4 inches ; breadth, below middle, $1\frac{5}{8}$. Those of the fossa on the 7th transverse process are—antero-posteriorly, at middle, $2\frac{3}{4}$ inches ; externally, 3 ; vertically, 3 ; depth, $\frac{1}{2}$ inch. The end of the 10th rib is $1\frac{7}{8}$ inch broad below the middle ; the 10th fossa, antero-posteriorly, $3\frac{1}{4}$ inches ; vertically, 4 ; depth, $\frac{5}{8}$ inch. But the edges of the fossæ are not completely ossified. The fossa faces downwards, and also outwards and backwards, the end of the rib upwards and inwards. The more sloping form of the ends of the ribs in Megaptera (from the less projection at the top) than in B. musculus corresponds to the more inwardly elongated direction of the fossæ in Megaptera. The articular ends are all broader in Megaptera than in B. musculus. No part of the costal fossæ in Megaptera appears to have been covered with cartilage.

The different form of the costo-transverse articulation in Megaptera, compared with B. musculus, has the result that there is more extensive contact of the opposing surfaces in Megaptera. The fossæ are more elongated vertically or transversely, and, spoon-like, receive a broader head. This difference may have reference to the much greater massiveness of the rib. But dissection of the ligaments and of these parts in their natural relation will be necessary for a complete explanation of the differences.

Sternum.

37. Table VII.—*Measurements of the Sternum*, given in inches.

	Megaptera.	B. musculus, 60 feet long.
1. Length,	$11\frac{1}{8}$	12
2. Breadth,	11	$18\frac{1}{4}$
3. Length of wing, transversely from where it joins the body,	$2\frac{1}{4}$	7
4. Breadth of wing, at the end,	$2\frac{3}{8}$	$4\frac{1}{2}$
5. ,, ,, at the middle,	$3\frac{1}{2}$	$3\frac{1}{2}$
6. ,, ,, where it joins body,	$4\frac{1}{2}$	$4\frac{1}{3}$
7. Length of cervical process,	4	3
8. Breadth of ditto at its base,	7	$5\frac{1}{4}$
9. ,, ,, at its middle,	$5\frac{1}{4}$	$3\frac{1}{3}$
10. Length of posterior process, from level of the wings,	$3\frac{1}{2}$	$5\frac{1}{4}$
11. Breadth of ditto at level of the wings,	6	5
12. ,, ,, at its middle,	$2\frac{1}{4}$	$2\frac{1}{2}$
13. Thickness of the beam between the wings, at middle line,	2	$1\frac{1}{2}$
14. Thickness at midway to end of wing,	$1\frac{1}{2}$	1
15. ,, at end of wing,	$1\frac{1}{4}$	$\frac{1}{2}$
16. ,, at middle of posterior process,	$1\frac{3}{4}$	$1\frac{1}{4}$
17. ,, of cervical process at middle,	$1\frac{1}{3}$	$\frac{1}{2}$
18. ,, of anterior edge at middle,	$\frac{3}{8}$	$\frac{1}{8}$
19. ,, ,, ,, at middle of side,	$\frac{1}{2}$	sharp
20. ,, at notch between wing and cervical process,	sharp	sharp
21. Depth of ant.-post. concavity of under surface,	1	1
22. Depth of transverse concavity of the upper surface,	$1\frac{1}{4}$	$1\frac{1}{2}$
23. Weight of the sternum, in ounces,	$31\frac{3}{8}$	$40\frac{1}{2}$

38. Characters in comparison with those of B. musculus.—The form of the sternum is shown in fig. 18.[1]

To understand the differences of form presented by the sternum in Megaptera and B. musculus it is necessary to con-

[1] For remarks on the interpretation of the sternum in Fin-Whales, the essential and non-essential parts, and the variation of the latter, with a drawing (fig. 4) showing its two places of articulation with the first rib, I may refer to my paper on B. musculus in this *Journal*, vol vi., November 1871. Reference may also be made to the figure given by Eschricht (*loc. cit.*, p. 139, fig. 47), showing the true relation of the sternum to the first ribs in a fœtal Megaptera, with his remarks on that point. Also to the figures of the sternum in B. musculus, B. borealis, and B. rostrata by Professor Flower (*Proc. Zool. Soc.*, 1864, p. 393). The sternum of the Fin-Whales is so liable to variation according to age and individual peculiarity, that care must be taken in attaching importance to the form presented by any individual specimen.

sider the very different proportions of the anterior aperture of thorax in these two species. This is seen if the first pair of ribs are laid at the natural distances from each other above and below. In **Megaptera** the vertical diameter of the ring (from the top of the ribs to the anterior border of their lower end) is 32 inches, the greatest transverse diameter 28 inches; in B. musculus the vertical diameter is 28 inches, the greatest transverse diameter 35. Thus in Megaptera the width is 4 inches less than the height, while in B. musculus the width is 7 inches greater than the height. The contrast is still greater on the lower part of the ring, the greatest width being above the middle. This difference appears to determine the form of the sternum in the two species. The sternum of B. musculus could not be fitted between the naturally-placed ribs of Megaptera; the outer ⅓ or ½ of the wing would have to lie upon the rib. The narrowness of the space between the first pair of ribs in Megaptera requires a short-winged sternum, and the more oblique course of the lower end of the rib is a reason why the hinder edge of the wing should slope to the posterior process instead of passing horizontally inwards. So in B. musculus the wide ring requires a wide-winged sternum, and the nearly horizontal direction of the hinder edge of the wing is in adaptation to the nearly horizontal direction of the anterior border of the lower end of the rib. These different relations also influence the position of the marks for the terminal costo-sternal articulation, and require the posterior process of the sternum to be longer in B. musculus.

The reasons for most of the differences seen in the table of measurements (Table VII.) of the sternum in the two species are now evident. Its somewhat diamond-shape in Megaptera is owing to the shortness of the wings, and to their sloping to a short posterior process; the form of a cross in the B. musculus, to the long wings and the long posterior process. The greater massiveness of the ribs in Megaptera requires the sternum to be thicker in it than in B. musculus.[1]

In Megaptera the *cervical process* is very broad, and is in an

[1] The weight of the first rib is—in Megaptera, right 182 ounces, left 168; in the B. musculus—right 110½ ounces, left 110. But the left rib in the B. musculus has a broader sternal end than the right.

M

unfinished condition, the subcartilaginous border, ⅜ inch thick at the middle, becoming narrower on the sides to within an inch of the base, where the edge becomes thin and completed. There is no indication of bifurcation at the front.[1] It is bent downwards very much at its base, this being the sole cause of the antero-posterior concavity of the inferior surface of the bone. On the upper surface the cervical process is rather suddenly bent downwards at the middle, rendering it convex, to which the forward thinning of the process also contributes. The moderate longitudinal convexity on the posterior half of the upper surface of the sternum is owing to the greater thickness of the beam between the wings, not to any bending down of the posterior process. It may be noted that the concavity on the border between the cervical process and the wing is not quite so deep on the left side as on the right.

39. RELATION OF THE STERNUM TO THE FIRST RIB.—The macerated bones alone would be very apt to mislead in the articulation of the skeleton. As Eschricht has shown with his usual

---

[1] The variation in this respect in the fin-whales is great. In my 35-feet-long *B. borealis*, the cervical process, after a length of 1½ inch, has bifurcated for half an inch, ossifying into a large undivided plate of cartilage, 2 inches in length by 4 in breadth. In the 50-feet-long *B. musculus* the cervical process, 3½ inches long, shows no appearance of past or coming bifurcation. In the figure given of *B. musculus* by Professor Flower (*loc. cit.*) the wings and cervical process form one great irregular semilunar plate with no sign of bifurcation, present or obliterated. In a sternum in my possession, found on the shore, which appears from its other characters to be that of *B. musculus*, the anterior process is very like that in the figure given by Professor Flower, except that there is a great median fissure, about 5 inches deep by 1 inch wide, nearly closed in front. In my figure of the 64-feet-long *B. musculus* (*loc. cit.*, 1871), the small foramen is so far back that it might be attributed to the position of a blood-vessel rather than to a former bifurcation of the bone. In two figures of the sternum of *B. musculus* given by Van Beneden and Gervais (*loc. cit.*, pl. xii.-xiii. figs. 14), one of them from a young animal, it is bifurcated anteriorly. The same authors give diagram figures of the sternum of *Megaptera* (*loc. cit.*, p. 128), showing a deep bifurcation closed in to form a foramen, and then the foramen obliterated. In their figure of Megaptera Lalandii (pl. ix. fig. 5) the bone is widely bifurcated for more than half of its entire length, and they speak of the sternum of Megaptera longimana and Megaptera Lalandii as "assez semblable" (p. 134). The sternum of this Megaptera would, to all appearance, never have bifurcated, nor is there any sign of a filled-up aperture. The smooth inferior surface shows the blood-vessel grooves and perforations radiating from the middle thick part on the posterior half of the bone (the perforations towards the thick part), outwards on the wings, and forwards, symmetrically and undisturbed, on the cervical process.

care, the rib lies behind the wing of the sternum. The adaptations in the developed **bones present** several points of interest.

[To understand these in Megaptera it is necessary to refer to the adaptations in *B. musculus*. As seen in my figure of the parts, still in their natural connection (*loc. cit.*, 1871), there are two places of articulation—(1) the *lateral costo-sternal joint*. The end of the wing, covered by cartilage, is joined on its posterior edge by ligaments to the anterior border of the rib, where a rough mark **is seen.** (2) *Terminal costo-sternal joint*. The end of the rib advances towards the posterior process of the sternum, and the short cartilage belonging to the anterior part of the end of the rib is joined to the sternum by ligament. The rough mark for this articulation is very evident in the 50-feet-long B. musculus, on the angle between the wing and posterior process, and on the anterior half of the side of the process, which is specially broadened at this part. The great length, transversely, of the wing of the sternum in B. musculus gives room for a considerable space between the lateral and terminal joints. Viewing **the end of** the **rib** in the 50-feet-long B. musculus, it is seen **to be** divided into three parts—an anterior, $1\frac{1}{2}$ inch thick, a narrow **middle** part, $\frac{1}{4}$ to $\frac{1}{2}$ inch thick ; and a posterior part, $\frac{3}{4}$ to 1 inch thick. In the 64-feet-long **B.** musculus the narrow middle part is **a sharp-**finished edge, the anterior and posterior parts broad, **and each covered** by cartilage. The anterior cartilage is at the terminal costo-sternal joint, the posterior cartilage is free and non-articular, and is a long way behind the sternum. The narrow middle part is the most projecting part of the rib, the free border receding from it, covered by the posterior cartilage.

In my *B. borealis* the soft parts here have been preserved, and throw light on the transition from B. musculus to Megaptera. Externally is seen the lateral costo-sternal articulation, as in B. musculus, the cartilage belonging to the sternum. Internally, after a short interval, owing to the shorter wing, the cartilage, which has been detached from the rib, is, like the end of the rib itself, 6 inches in breadth, and almost separated by a narrow middle part into an anterior and posterior part. The anterior and smaller part, $1\frac{1}{2}$ inch **in** length, close to the sternum, is joined to it by ligament, to the **angle** formed by the wing and a special projection from the side of **the base of** the short posterior process ; the posterior part, $2\frac{1}{2}$ inches **in length, is** joined to the fibrous membrane which fills up the great gap (5 **inches** deep) behind the sternum, and between the greater part **of** the free ends of the first pair of ribs. The end of the rib recedes **behind** the narrow middle part, but the cartilage that covers it projects internally, to **the** extent of $1\frac{1}{2}$ inch posteriorly, and **thus** gives the whole cartilage **of the** end of the rib a deeply notched or scooped-out form, obliquely, **resembling** the scooped-out end of **the** long **rib** in Megaptera.]

There is no evidence of the presence of two joints in Megaptera, but rather of one continuous oblique ligamentous connection. The end of the wing of the sternum has been covered by cartilage; the surface, $2\frac{5}{8}$ inches by $1\frac{1}{4}$, facing outwards, with a little obliquity upwards and backwards. The rough mark for the ligamentous attachment, seen best on the left side, is on the oblique posterior border of the bone, formed by the hinder edge of the wing and the side of the broad base of the posterior process. It is $4\frac{1}{2}$ inches in length, $\frac{3}{4}$ inch broad externally, narrowing gradually to $\frac{1}{8}$ inch at the inner end. It has not been covered by cartilage. In contrast with the nearly rectangular recess between the wing and the posterior process in B. musculus, this part in Megaptera is thus filled up, giving an oblique posterior border to the bone, and contributing to its diamond form. The ligamentous mark does not extend back upon the narrow part of the posterior process, on which a sharp edge separates, for the last $2\frac{1}{2}$ inches, the flat upper surface from the convex side of the process. But the ossification here is not complete on the right side. On the right side of the upper surface there is a sub-cartilaginous area, $3\frac{3}{4}$ inches in length by $\frac{3}{4}$ inch broad, extending to the very point. Further ossification of this cartilage backwards might have elongated the process, and further outward growth of its anterior part, and of the neighbouring part of the wing, would have filled up the border on the right side to the same extent as on the left. The partial concavity or oblique recess behind the wing on the right side corresponds to the presence of a greater projection on the anterior angle of the rib on the right than on the left side.

[In the *B. musculus* the rough mark for the terminal joint extends for $2\frac{1}{2}$ inches on the wing and for $3\frac{1}{2}$ on the posterior process, to the part where the process becomes considerably narrower, 2 inches from the point. The part of the mark situated on the process is much broader ($1\frac{1}{4}$ inch) as well as longer than the part on the wing. On the left side the side of the process is quite flattened by the mark, and the recess between the process and the wing is deeper than on the right side. The asymmetry here may be related to the greater breadth (by nearly 1 inch) of the left than the right rib, at the end. On the hinder border of the wing there is an interval of $2\frac{1}{2}$ inches between the mark for the internal joint and where the outer end of the wing begins to be sub-cartilaginous. The interval, however, is rough, as if it had attached a ligament. The first rib shows a distinct rough elevation on its anterior border, for the lateral costo-sternal articulation, beginning 4 to 5 inches from the end of the rib and occupying 3 to 4 inches. This corresponds to the hinder slope of the outer end of the wing of the sternum.]

In Megaptera there is no separated mark on the rib indicating a distinct lateral costo-sternal articulation, simply the border of the rib is thick and rough for its last 7 to 8 inches, in marked contrast to the character of the border of the rest of the rib. This rough part extends 2 to 3 inches farther up the rib than the level of the fore part of the end of the wing of the sternum, as placed in fig. 18.

This border of the rib, on the right side, terminates in a prominent angle or process, the end of which shows a sub-cartilaginous surface, $1\frac{1}{2}$ inch by 1 inch. The mode of articulation I infer to have been by continuous ligamentous attachment, externally, to the last 7 or 8 inches of the rib and to the cartilage on the terminal process of this border of the rib; internally, to the cartilage of the wing of the sternum and to the rough mark on the hinder border of the wing and side of the base of the posterior process. On the left rib there is no terminal process to the anterior border of the rib, simply a rounded-off angle, without thickening, but it appears to have been covered by cartilage. As shown in fig. 18, the left first rib is shorter at the lower end than the right by $1\frac{1}{4}$ inch.

Whether the wide oblique notch in the end of the first rib is exceptional in this Megaptera I have not the means of determining. The edge is thin, and as if finished on the right rib, but on the left a thin strip of cartilage appears to have been continued on the anterior part of the notch. The notch, or excavation, is on the right rib 4 inches in length, $\frac{3}{4}$ inch in depth; on the left rib 5 inches in length, $\frac{2}{3}$ in depth. The projecting end of the rib behind the notch is 3 inches in breadth, 1 inch in thickness towards the posterior end, and has been covered with cartilage. The end of the rib, therefore, has, as in the adult B. musculus, two cartilages—the anterior, by which it articulates with the sternum; the posterior, a long way behind the sternum. Between these, in the adult B. musculus, is a thin finished edge of bone. In Megaptera this thin part is still less developed, so that there is a wide notch between the parts bearing the posterior and the anterior cartilages. But, further, in Megaptera the rib stops short on its anterior border several inches external to what would be its end in B. musculus, giving the notch its outward obliquity. Prolongation of this border would have carried it upon or behind the point of the posterior process of the sternum. It therefore stops short when it arrives below the wing. Thus the narrowness of the anterior aperture of the thorax and the sloping direction of the rib account for these modifications of the end of the first pair of ribs, as well as for the form of the wing and posterior parts of the sternum, and for the difference in the mode of articulation, as compared with B. musculus.

[The very different proportions of the sternum in the *B. musculus*, compared with Megaptera, are seen in the measurements in the table (Table VII.). The wings are very long transversely, and more expanded near the end than at their middle ; the posterior process is long and bent downwards, more than the cervical process is. The cervical process is much narrower than in Megaptera. The whole bone is thinner. The parts of the edges the ossification of which is unfinished, are, the outer edge of the wing, $\frac{1}{4}$ to $\frac{1}{2}$ inch thick (the posterior slope of the end thicker than the anterior, the surface facing obliquely upwards) ; part of the point of the posterior process ; and parts of the anterior convex border of the cervical process, thinnest ($\frac{1}{12}$ inch) at the middle (the most projecting part), a little thicker ($\frac{1}{8}$ inch) on each side. Along the posterior $\frac{2}{3}$ of the cervical process the border is quite sharp and finished. It is not evident how this sternum could bifurcate forwards, or how their thin edges could grow so as to fill up the great hollow between the wing and the cervical process.]

The Chevron Bones.

40. CHARACTERS IN COMPARISON WITH THOSE OF B. MUSCULUS.—The chevron bones in this Megaptera are 10 in number, the first and the two last in separate halves. The number in the B. musculus is 13, in the B. borealis 15. It is easy to distinguish those of Megaptera from those of the other two finners. As seen in the table of measurements given below (Table VIII.), the arch in all of them in Megaptera is much wider at the middle than at the top, the space having the form of the lower $\frac{3}{8}$ of an ellipse or of a rather pointed oval.

[In *B. musculus* the space in the first three is a wide triangle, 4 inches deep, $3\frac{1}{2}$ across, the laminæ not coming at all towards each other at the top. On the 4th, they begin to approach a little at the top, rendering the space a little narrower there than a short way below, and this increases backwards, but not so much as in Megaptera

until we reach the 11th and 12th of B. musculus, which in this respect closely resemble the 7th and 8th of Megaptera.]

This approximation of the lamina above results from the
greater breadth of the *articular surface* in Megaptera. The
measurements of this surface on the 4th are, in Megaptera,
breadth 2 inches, length 5; in B. musculus, breadth 1⅔ inches,
length, 4½. The surface in Megaptera is elliptical, with the
greatest bulge on the inner side. It is bevelled before and
behind so as to present two facets; the anterior rests on the
vertebra to which the chevron bone belongs, and is the more
flattened of the two; the posterior rests on the intervertebral
disc; in Megaptera it could not reach the vertebra behind,
owing to the great length of the discs.

[In *B. musculus* the articular surface, besides being narrower, has
a different form, the inner side concave on the three first, nearly
straight on the next two, convex behind on the next three, and on
the four posterior the surface becomes elliptical but not so broad as
in Megaptera. The posterior facet, separated from the anterior by
a middle rounded part, may rest on the vertebra behind. The *arch*
(included space) is about as large as in Megaptera.]

The *spines* are but little developed in height or in breadth
(antero-posteriorly) compared with those of B. musculus. This
is, at least in part, owing to the less advanced ossification in
Megaptera, but the lower edge of the chevron spines are unfinished in both. This edge has very little of the convexity
which is so marked nearly all along the series in B. musculus,
giving them the semicircular form in the latter. The 2nd
alone in Megaptera shows much convexity, the 3rd and 7th a
little; the 4th is, on the whole, the best developed of the
chevron spines. In the subjoined table of measurements of
the chevron bones of Megaptera, the 4th of B. musculus is
given for comparison.

Individual Chevron Bones.—The *first* presents two separate
triangular laminæ, 3½ inches in height, about 3 inches broad at
the top, the point directed downwards and forwards. The *ninth* is
smaller than the 1st, its more blunt point directed straight down.
The laminæ of the *tenth* are oval, 1¼ inch in height, 1¾ antero-
posteriorly. Their articular surface, however, is very distinctly
marked above, and reaches some way down on the inner side,
very sharply marked off from the smooth oval internal surface

proper.　The *seventh* shows exceptional narrowness of the articular surfaces (breadth, 1¼ inch; breadth of sixth, 2 inches; of eighth, 1¾ inch), with consequent thinning of the laminæ and much greater width of the arch than in the sixth.　This is the chevron bone belonging to the first vertebra, in which the anterior and posterior hæmal tubercles meet to form a continuous ridge.

[The corresponding chevron bone in *B. musculus* (the 12th) is that on which the articular surface begins to diminish to a marked extent in breadth, but the narrowness is continued on the chevron behind it.]

41. TABLE VIII.—*Measurements of the Chevron Bones of Megaptera, and of the 4th Chevron Bone of B. musculus, in inches.*

	Megaptera.										B. musc.
	1	2	3	4	5	6	7	8	9	10	4th
Height of the arch,	..	3	3	3	2⅝	2⅝	2¼	1¾	...	...	4½
Width of arch at top,	...	1¼	1	1	1	1	1¾	1	...	...	3¼
,,　　,, at the middle,	...	1½	1⅝	1⅝	1½	1½	2	1½	...	...	2½
Breadth of lamina at top,	3¼	4⅞	5	5	5	4½	4⅞	3½	2¾	1¾	4¼
,,　　,, at junction,	...	3	3¾	3¾	4	3¾	3¾	3	...	...	3½
Breadth of spine (ant.-post.),	..	3½	5	5½	5½	4	4½	3	...	...	5¼
Height of spine,	...	3⅞	3½	3	2½	2	2¼	1½	...	...	5¼
Height of entire bone,	3½	7	7	6½	5⅝	5⅞	5	3½	2¾	1¼	10

42. EXPLANATION OF PLATE VI.

Fig. 17.　Posterior aspect of atlas of Megaptera longimana, reduced to ⅛.　*a, a,* the lateral articular surfaces; *b,* mesial articular surface; *c,* ligamentous area.　The form of the spinal canal, its neural and odontoid parts, is seen.　Compare with this figure the figure showing the same aspect of the atlas of B. musculus, with the transverse ligament, in this *Journal,* vol. vii., 1872, fig. 5.　In Megaptera observe especially the quadrate form of the ligamentous area, and the presence of a mesial articular surface separating the lateral articular surfaces below.　The difference between the right and left shallow groove between the lateral and mesial articular surfaces is seen in the figure.　The small surface seen below the mesial articular surface, on the subaxial peak of the atlas, is not articular.　The very slight projection of the lateral articular surfaces at the transverse processes is seen in the figure.

Fig. 18.　Sternum and first pair of ribs of Megaptera longimana, seen from before, reduced to 1/12.　*a, a,* the rough part on the inner border of the rib near the wing of the sternum ; b, b', anterior angle of right and left rib, less developed on the left ; c, c', posterior angle of the ribs. The left rib is seen to be shorter than the right. The notch between the two angles here is seen to be shallower on the left rib.

Sternum.—The dotted part on the cervical process is where it has been covered by cartilage. The sub-cartilaginous end of the wings is not seen when viewed from before. The figure shows the more horizontal direction of the posterior border of the right wing, in adaptation to the greater projection of the anterior angle of the right rib. On the left side, the hollow between the wing and the cervical process is seen to be more filled up than on the right side, but the difference on the two sides is not so great as that between the wing and the posterior process. The radiating lines indicate the direction of the vascular grooves and foramina, converging to the thicker part of the bone.

Compare with this figure the figure of the corresponding parts in B. musculus, in this *Journal*, vol. vi., 1871, fig. 4, showing the ligaments of the two costo-sternal articulations, &c.

PART IV.

THE SKULL.

The following table of measurements, besides showing the size and proportions of the various parts of the skull of this Megaptera, will enable these to be compared with the corresponding parts of B. musculus. The measurements to which an * are prefixed are those given in the same order by Professor Flower (*Proc. Zool. Soc.*, 1864, p. 411) of several skulls of B. musculus, with which the sizes in the second column of my table may be compared. Some of the measurements given in the

N

1. TABLE I.—*Measurements of the Skull,* given in inches.

	Megaptera, 40 feet long.	B. musculus, 50 feet long.
*1. Length of skull, in straight line, to point of upper jaw, ...	125	145
*2. Breadth of condyles, ...	$12\frac{1}{4}$	$11\frac{3}{4}$
3. Condyle, height, ...	$9\frac{1}{4}$	10
4. ,, breadth, ...	$5\frac{1}{4}$	$5\frac{3}{8}$
5. Projection of condyles behind most projecting part of exoccipitals, ...	$2\frac{1}{4}$	$1\frac{1}{4}$
6. Projection of condyles behind bone immediately external to them, ...	3	4
7. Depth of depression between condyles and between posterior part of exoccipitals, .	$2\frac{1}{4}$	$3\frac{1}{2}$
8. Depth of supra-occipital depression, . .	$3\frac{1}{2}$	$2\frac{3}{4}$
9. Foramen magnum, height, ...	$4\frac{7}{8}$	$4\frac{1}{2}$
10. ,, width, ...	$4\frac{5}{8}$	4
*11. Breadth of exoccipitals, ...	$41\frac{1}{2}$	43
12. Descent of exoccipitals below level of condyles, .	$3\frac{1}{2}$	2
13. From exoccipital to outer edge of squamosal, horizontally out, ...	15	12
14. Descent of squamosals below level of condyles,	14	10
*15. Breadth of skull (greatest) at zygoma of squamosals, ...	$71\frac{1}{4}$	$66\frac{1}{2}$
16. Ditto near posterior angle of zygoma, . .	60	58
*17. Length of supra-occipital, ...	29	31
18. Length of articular process of squamosal (zygoma), ...	$22\frac{1}{2}$	$27\frac{1}{2}$
19. Malar, length of outer border, straight, .	8	$8\frac{1}{4}$
20. Ditto, breadth at narrowest part, . .	$2\frac{5}{8}$	$1\frac{1}{2}$
21. Outer edge of orbit, length, ...	9	9
22. ,, ,, height, ...	8	7
23. Temporal passage, width at front, . . .	20	19
24. ,, antero-posteriorly, greatest,	$8\frac{1}{4}$	$10\frac{1}{2}$
25. Temporal fossa, transversely, greatest, . .	31	28
26. ,, antero-posteriorly, . .	34	42
27. ,, greatest depth, . . .	$10\frac{1}{2}$	$10\frac{3}{4}$
28. ,, anterior edge anterior to peak of nasals, . . .	$\frac{1}{2}$	$10\frac{1}{2}$
29. ,, posterior edge to posterior border, ...		
30. ,, of squamosal (sagittal), .	18	15
*31. Orbital plate of frontal, transversely, greatest,	31	27
*32. ,, ant-post., at inner part, . .	27	29
*33. ,, ,, at outer part, upper surface, . . .	12	$15\frac{1}{2}$
34. ,, supra-orbital edge, straight, .	8	10
*35. Nasals, length, inner border, ...	$9\frac{1}{2}$	$7\frac{1}{2}$
36. ,, ,, outer border, ...	$9\frac{5}{8}$	$10\frac{3}{4}$
*37. ,, breadth of the two, at posterior end, .	$4\frac{1}{4}$	$5\frac{7}{8}$
*38. ,, ,, at anterior end of outer border, .	$10\frac{1}{4}$	$10\frac{1}{2}$
39. ,, ,, at where inner border ceases, .	$10\frac{1}{4}$	$8\frac{1}{2}$
40, 41. ,, breadth of each at middle of inner border, upper surface, . .	$1\frac{1}{2}$	$3\frac{1}{4}$

TABLE I.—*Measurements of the Skull*—continued.

	Megaptera, 40 feet long.	B. musculus, 50 feet long.
42. Nasals, greatest thickness, between upper and under surfaces, . . .	8	9
43. ,, width of depression between nasals, .	$1\frac{1}{2}$	$\frac{7}{8}$
44. Smallest breadth between temporal fossæ, across nasals, ...	$10\frac{2}{3}$	$15\frac{1}{2}$
45. Length of cranium, straight from end of occipital condyle to anterior end of nasal, .	39	38
46. Ditto, ditto, to anterior edge of temporal fossa,	40	49
47. Ditto, ditto, to transverse frontal fossa, . .	32	$32\frac{1}{2}$
*48. Length of beak, from anterior edge of temporal fossa, ...	85	96
49. Ditto, from transverse frontal fossa, .	94	$112\frac{1}{2}$
*50. Projection of premaxillary beyond maxillary, .	6	8
51. Length of maxillary, from end of frontal process, ...	$89\frac{1}{2}$	108
52. Ditto, from curved border at temporal fossa, .	$78\frac{1}{2}$	89
53. Ditto, from back part of ant.-orbital angle, .	94	102
54. Ditto, on palatal surface, . . .	96	107
*55. Breadth of maxillaries at hinder end, . .	18	17
*56. Ditto, across orbital processes, on the curve, .	80	70
*57. Breadth of beak at base, on the curve (just before anterior edge of temporal fossa), .	45	48
58. Ditto, straight, by callipers, . . .	$38\frac{1}{4}$	40
*59. Breadth of maxillary at same, . .	16	$16\frac{3}{4}$
*60. ,, of premaxillary at same, .	$1\frac{1}{2}$	0
61. ,, of mesial gap at same, . .	$9\frac{1}{2}$	$13\frac{1}{4}$
*62. Breadth of beak, $\frac{1}{4}$ of its length from the base, on the curve, . . .	$33\frac{1}{2}$	36
63. Ditto, straight, by callipers, . . .	$31\frac{1}{4}$	$33\frac{2}{3}$
*64. Breadth of maxillary at same, . . .	$8\frac{1}{2}$	12
*65. ,, of premaxillary at same, .	4	$4\frac{3}{4}$
66. ,, of mesial gap at same, . .	$8\frac{1}{4}$	$2\frac{1}{4}$
*67. Breadth of beak at middle, on the curve, .	$27\frac{1}{2}$	27
68. Ditto, straight, by callipers, . . .	$26\frac{1}{4}$	26
*69. Breadth of maxillary at same, . . .	$8\frac{1}{4}$	$8\frac{1}{2}$
*70. ,, of premaxillary at same, .	$4\frac{1}{4}$	$4\frac{3}{4}$
71. ,, of mesial gap at same, .	3	$1\frac{1}{2}$
*72. Breadth of beak at $\frac{3}{4}$ of its length from base, on the curve, ...	$19\frac{1}{4}$	17
73. Ditto, straight, by callipers, . . .	$18\frac{1}{2}$	$16\frac{1}{2}$
*74. Breadth of maxillary at same, . . .	$4\frac{1}{2}$	$3\frac{1}{2}$
*75. ,, of premaxillary at same, .	$3\frac{3}{4}$	4
*76. ,, of mesial gap, . . .	$2\frac{1}{2}$	$1\frac{3}{4}$
77. Premaxillary, length, ...	92	113
78. ,, breadth where maxillary ceases distally, . . .	$3\frac{5}{8}$	$3\frac{1}{2}$
79. Prenasal gap, greatest width, . . .	$14\frac{1}{2}$	$12\frac{1}{2}$
80. ,, length, from anterior border of nasals,	$31\frac{1}{2}$	25
81. Length of narrow intermaxillary space, from prenasal gap to end of beak, . .	55	84
82. Vomer, length, ...	90	121
83. ,, greatest depth of its cavity, . .	$8\frac{3}{4}$	$8\frac{1}{2}$
84. ,, greatest width of its cavity, . .	$7\frac{7}{8}$	$6\frac{7}{8}$

TABLE I.—*Measurements of the Skull*—continued.

	Megaptera, 40 feet long.	B. musculus, 50 feet long.
85. Palate bone, length, ...	24	25¼
86. ,, greatest breadth, . . .	10½	10½
87. Maxillary, on palatal aspect, breadth at base, straight, ...	20½	21
88. Maxillary, greatest depth of concavity there, .	5¾	8¼
89. ,, breadth at ¼ of its length from base,	18	18
90. ,, greatest depth of concavity there, .	4½	6½
91. ,, breadth at middle, . . .	14	14
92. ,, greatest depth of concavity there, .	3⅛	3⅝
93. ,, breadth at ¾ of its length from base,	9¾	9¼
94. ,, greatest depth of concavity there, .	1⅜	1⅜
95. Cranial cavity, length from lower edge of foramen magnum to edge of olfactory fossa, . . .	12	16
96. ,, greatest breadth (at temporal fossa), . . .	18	16
97. ,, greatest height (at sella turcica), . . .	8	8
98. Olfactory fossa, length, . . .	4	5
99. ,, breadth, at entrance, . .	3	3¼
100. ,, height, at entrance, . .	2	3

table may seem unnecessary, but they may be useful in the study of the skull in fin-whales. In making the measurements and the observations which follow, I had the skulls of the Megaptera and the B. musculus placed together, so that they could be approached on all sides and studied comparatively.

2. LENGTH AND PROPORTIONS OF THE SKULL.—The length of the skull, taken to the point of the upper jaw, is less than that given as the length of the head (Table I. Part III.), by the amount of projection of the mandible beyond the beak and of the soft parts in front of the mandible. The amount to which the condyles project behind the rest of the skull falls to be deducted in estimating how much the head forms of the total length. The comparison of the parts in the two skulls in the following observations is between the actual sizes, but it is to be borne in mind that the Megaptera was 40 feet long, the skull 125 inches in length, the B. musculus 50 feet long, the skull 145 inches in length. In the true comparison of the lengths of the several parts of the skulls, therefore, about a seventh has to be deducted from those of B. musculus, or a sixth added to

those of Megaptera. In like manner the comparison of the actual breadths does not bring out the full proportionate breadths in Megaptera, unless it is borne in mind that the skull of Megaptera was only 125 inches in length and that of B. musculus 145.

3. OCCIPITAL BONE AND REGION.—The *foramen magnum* is larger in Megaptera than in B. musculus, and broader in proportion to its height (see measurements in the Table). At the upper part of the foramen the bone is thin in Megaptera (not a ¼ inch), in B. musculus very thick (2 inches).

Condyles.—The condyle is narrower in Megaptera, but the two condyles together are wider in Megaptera owing to the greater width of the intercondyloid space. In B. musculus the lower end of the foramen magnum joins a narrow intercondyloid triangle (1 inch wide) at an acute angle; in Megaptera the angle is obtuse, and the upper part of the intercondyloid space is a wide triangle (3 inches wide, 2 in length). The space below this is not roofed over in Megaptera, but is' so in B. musculus. In Megaptera the condyles begin above by a non-articular ridge 2 inches in length, beginning 1 inch above the foramen; in B. musculus the articular condyle begins at once, and on a level with the top of the foramen. The condyle is more bent vertically in Megaptera. The *projection of the condyles* beyond the rest of the back of the skull [1] is greatest in Megaptera, but their projection from the level of the bone immediately external to them is 1 inch more in B. musculus than in Megaptera. This will allow greater freedom and extent of motion in B. musculus.

Supra-occipital.—In B. musculus there is a very well-marked sharp median ridge, rising to an inch in height, ending behind in a sharp diamond-shaped tuberosity, 7 inches from the foramen magnum. This ridge is very obscurely marked in Megaptera, which presents a broad deep median hollow along the whole length of the bone, bounded on each side by the smoothly convex side of the bone. In B. musculus these parts have irregular muscular markings, and the hollow is less. The upper end of the supra-occipital, at the transverse frontal fossa, is narrower in Megaptera (10 inches) than in B. musculus (14 inches).

[1] This projection does not occur at all in B. borealis or in B. rostrata.

The *ex-occipitals* are not so broad in Megaptera as in B. musculus, and are smoothly convex in both directions from a little way external to the condyles; the most projecting part is at the junction of the outer and middle thirds, from which the surface falls smoothly to the outer edge, without any par-occipital ridge. In B. musculus there is a transverse depression along the outer half, the surface more sharply bent vertically and flatter transversely; the most projecting part is at the extreme outer edge, rising into a prominent par-occipital ridge. The ex-occipitals descend below the level of the condyles more in Megaptera (3½ inches) than in B. musculus (2 inches).

Basi-occipital.—In Megaptera the sub-occipital notch between the inner walls of the ear-bone spaces, and receiving the vomer, is wider in Megaptera (15 inches, in B. musculus 13). This part will be further noticed with the vomer. External to this, the sharp triangular fissure in the occipital margin, opening into the ear-bone space, is much deeper and wider in Megaptera (depth 4 to 5 inches in Megaptera, in B. musculus 2 inches).

Occipital Plate of the Temporal Bone.—This plate, in its form and great size, presents one of the most striking differences between Megaptera and B. musculus. Its outer and lower part (post-mandibular process) has the glenoid fossa in front for the support of the condyle of the mandible. It is much larger and more square-shaped in Megaptera. The measurements are, along lower border, sloping downwards and outwards, in Megaptera 13 inches, in B. musculus 12; transversely at about the middle, outer border moderately concave, Megaptera 15, B. musculus 12; transversely at posterior angle of temporal fossa, in Megaptera 18½, in B. musculus 13½; antero-posteriorly, at inner part, in Megaptera 18, in B. musculus 14½; at outer border, in Megaptera 23 (going as far as the articulation with the frontal bone and to four inches from the anterior end of the zygoma), in B. musculus, as seen from behind, also about 23, but indefinite from the rounding off into the long zygoma, and about 11 inches from the anterior end of the zygoma. The outward slope of this border is such that its upper part is more external than its lower, in Megaptera 7 inches, in B. musculus 2½. The upper border, going outwards, **slopes in** Megaptera a **little** downwards, in B. musculus **very** much upwards.   **The more** square form in Megaptera is mainly owing to the greater height at the inner part, **from the temporal** fossa cutting less back into it, and to the greater general breadth.

4. TEMPORAL FOSSA AND PASSAGE.—When these skulls are viewed anatomically, from above as well as from the side, it is seen that the parts connected with the temporal muscle present the most marked of all the characters which differentiate the skull of Megaptera.   Considering the enormous length of the mandible in whalebone whales, and its weight even in the water, we would expect to find a correspondingly large temporal fossa and passage, and, as the head is flat, that the fossa would assume an expanded form.

From the measurements given in the table, it is seen that the *temporal fossa* is broader by 3 inches in the Megaptera, and longer by 8 inches in B. musculus, and is of nearly the same depth in each, at the deepest part, which is towards the back part of the inner side.   From 5 to 6 inches of the greater length in B. musculus is obtained by an angular prolongation behind where the zygoma and squame of the temporal meet, while in Megaptera the posterior border is uniformly concave. The great difference, however, is in the shortness of the fossa at its outer compared with its inner part, being in Megaptera scarcely half as long externally as it is internally.   This is owing to the well-known obliquity backwards of the orbital process of the maxillary and of the anterior border of the orbital plate of the frontal, so that at the outer part that plate is but 12 inches in length, as compared with 27 internally, two of the 27 covered by the maxillary at the anterior end of the fossa.   In B. musculus the obliquity is very much less, the measurements at the corresponding parts being respectively 15½ and 29 inches; but about 4 inches of this narrowing is obtained by a forward direction of the posterior border of the plate, while the posterior border of the plate in Megaptera is, on the contrary, directed a little backwards.   The less length of the temporal fossa, and its farther back position on the skull, in Megaptera are made manifest by observing its relation to the ends of the maxillary and nasal bones.   A line

drawn across the skull where the maxillary and supra-occipital
bones come in near relation (the **transverse** frontal fossa), has
in front of it, in Megaptera less than a third of the length of
the temporal fossa (9 to 10 inches), only its narrow anterior
end, and passes at about 3 to 4 inches in front of the orbit;
while in B. musculus well on to half of the fossa (18 inches) is
in front of the line, and it passes across the posterior end of the
orbit. In relation to the nasal bones, the anterior boundary of
the temporal fossa reaches, in Megaptera to only ½ inch in
front of the median peak of their anterior ends, in B. musculus
to 10½ inches. A difference is seen also in the level of the
floor of the fossa, especially on the outer half of the frontal
plate, which in Megaptera is convex with a general fall backwards (except at the anterior angle), while in B. musculus it is
flat with a slight fall forwards. Standing behind the skull,
with the eye on a level with the upper border of the occipital
plate of the temporal bone, there is seen above it, in B. musculus only the inner part of the anterior end of the temporal
fossa; in Megaptera the fossa, with its great slope, is seen in
its whole length, and the outer half of the posterior border is
also seen.

The differences in the temporal fossa in Megaptera and B.
musculus appear to be mainly owing to the greater breadth of
the cranium in Megaptera, in adaptation to a wider mandible
requiring the post-mandibular plate of the occipital to be not
only farther out but also farther back.

Temporal Passage.—This large aperture, leading down from
the back part of the temporal fossa upon the condyle or neck
of the mandible, is a continuation of the temporal fossa, but
may be conveniently distinguished by the above name. It is
large enough to allow a man's body to pass. In Megaptera the
posterior wall is uniformly and smoothly concave, with a depth
of 10 inches at the middle, and but faintly marked off from
the glenoid surface below, both being nearly vertical. In B.
musculus there is a well-marked angle in the posterior wall;
the depth at the middle is 7 inches, and the wall here is
marked off from the glenoid surface by a strongly pronounced
transverse ridge, towards which both surfaces slope forwards.
The angle on the posterior wall gives the passage a triangular

form in B. musculus contrasting with the semilunar form in Megaptera.

The *anterior boundary* of the temporal passage is formed by the thick **posterior border of the** supra-orbital plate of the frontal bone, rounded by the bone turning downwards and forwards like a scroll, bounding the orbital cone posteriorly, and forming a pulley-surface for the play of the temporal muscle. In Megaptera it is directed a little backwards, with slight concavity; in B. musculus it slopes forwards in its whole length. (Thickness at the inner, middle, and outer parts, in Megaptera 6, 3½, and 2 inches; in B. musculus, 5, 3, and 1½.) The direction of this great post-orbital bar is determined by, or determines, the direction of the orbit.[1]

5. DIFFERENCES OF THE BONES FORMING THE TEMPORAL FOSSA AND PASSAGE.—The bones to be noticed here are the parietal, at the fossa and passage, and behind and below the orbital cone; at the inner part of the passage, the temporal, sphenoid, and pterygoid; below the passage, the pterygoid, temporal, and palate; and, behind and below the orbital cone, the pterygoid as well as the parietal.

The *parietal* bone has much greater expansion on the fossa in Megaptera than in B. musculus, covers the inner wall as far forwards as to about 3 to 4 inches from the anterior edge of the fossa, and extends outwards on the floor to a breadth of from 6 inches at the middle of the fossa to 8 inches at the bar, and out upon the back of the bar for 13 inches as a thin lamina. In B. musculus it covers only the inner wall, forwards to within about 6 inches of the anterior edge of the fossa, none of the floor, and does not reach out on the bar. Where the bar is covered by the parietal in Megaptera, it is rough in B. musculus, along fully its inner half, and was in the recent state covered by cartilage. The parietal is seen to cease by a natural edge 1 inch internal to this rough part. On the posterior wall of the temporal fossa and passage, the parietal goes outwards farther

---

[1] In this connection it is interesting to note the different direction of this great bar in the different Finners. Its direction in Megaptera and in B. musculus is noted above. In B. borealis it is directed considerably backwards, giving the supra-orbital plate great breadth externally and its square form. In B. rostrata still more backwards, giving a rhomboid plate.

in B. musculus than in Megaptera (10 inches, in Megaptera
7 to 8); the descending parieto-temporal suture curves differ-
ently, its outward projection in B. musculus much greater above
than below, in Megaptera greater below than above; the border
then sweeping forwards in B. musculus to reach the level of the
top of the sphenoid, in Megaptera down to the level of the top
of the pterygoid.

The *temporal* bone here is divided for some way into two
parts by a suture-like fissure, passing upwards and backwards.
The lower part, much the broadest, articulates with the pterygoid in both to about the same extent, but is smaller above
in B. musculus than in Megaptera (3 inches at the middle,
Megaptera $4\frac{1}{2}$), and the fissure in B. musculus runs into the
angular recess of the wall and then bends sharply down. The
form of the upper part is determined by the relation to the
parietal; in Megaptera broad above, tapering below like a
sickle, its point just reaching the top of the pterygoid; in
B. musculus its continuation downwards is as broad as the hand,
reaching forwards to articulate with the pterygoid for 3 inches
and with the base of the sphenoid wedge.

The *sphenoid* bone shows itself on the surface here in
B. musculus, but not in Megaptera. In B. musculus it has the
form of a narrow wedge, 6 inches in length; height of base,
posteriorly, 1 inch, tapering forwards to a sharp point at the root
of the orbital cone. It articulates posteriorly with the temporal, above with the parietal, below with the pterygoid. It
and the parietal are seen to end at the same point together at
the root of the orbital cone. The parietal is thus cut off from
reaching any part of the pterygoid, by the interposition of the
sphenoid, in contrast with the condition in Megaptera.[1]

---

[1] The sphenoid shows itself here in B. borealis and also in B. rostrata, but
does not cut off the parietal from meeting the pterygoid. In *B. borealis* it is
wedge-shaped, shortened behind, so that the parietal meets the pterygoid behind
it for 2 inches. The temporal is thus cut off from meeting the sphenoid, but its
part above the fissure meets the pterygoid for $1\frac{1}{2}$ inch. In *B. rostrata* the
sphenoid here is 1 inch in length on the right side, $1\frac{1}{2}$ on the left, and about
$\frac{1}{2}$ inch in height, the wedge shortened in front so that the parietal meets the
pterygoid in front of the sphenoid for $1\frac{1}{4}$ to $1\frac{1}{2}$ inch. The tongue-like process of
the temporal meets the posterior end of the sphenoid, and below that the
pterygoid. It might seem that the concealment, or exclusion, of this part of the
post-sphenoid here in Megaptera was owing to the parietal having to reach out on

The *pterygoid* bone, as seen here, is in B. musculus square shaped, 6 inches in height, 4 in breadth, directed upwards and backwards; in Megaptera rather triangular, 4 inches in height, 2 to 2½ inches in breadth, directed very obliquely backwards and upwards. In Megaptera it is only on the right side that its narrow apex is actually touched by the sickle-like part of the temporal, on the left side there is an interval of ¾ inch. In Megaptera this temporal aspect of the pterygoid is marked off sharply, at about a right angle, from the basilar aspect, while in B. musculus the surface is continued down in a much more rounded form. Here the lower division of the temporal is seen to send forwards a process across the pterygoid, just below the root of the orbital pedicle, in Megaptera to within about 1 inch from the palate bone; in B. musculus the interval, at the narrowest, at the foramen, is 2½ inches. The greater convexity of the pterygoid bone here, in B. musculus, is owing to the bulging of the wall of the auditory space. The pterygo-temporal foramen here (admitting two fingers) is modified accordingly, elliptical and oblique in Megaptera (3 inches by nearly 1 inch), the pterygoid bounding only the anterior end; in B. musculus, ovoid, the pterygoid forming the inner half.

The *palate* bone at this region is different. In Megaptera, where the posterior and upper borders meet, it sends up a tri-angular process to below the root of the orbital pedicle, pushing the pterygoid outwards; upper border of palate bone deeply concave for the first six inches; posterior border concave back-wards; foramen between palate, pterygoid, and frontal admits a finger. In B. musculus the palate bone here wants all these characters; the posterior edge is very convex, indeed has a blunt angle above the middle. The foramen is merged in the long

the pedicle of the orbital cone, but in B. borealis the parietal reaches out below the cone for 3 inches beyond the first inch occupied by the pterygoid, and far enough forwards to cover the fissure between the two scrolls. Again, that it is not the backward direction of the post-orbital bar that determines the extension of the parietal upon it in Megaptera, is seen by the fact that the backward direction of the bar is greater in B. borealis than in Megaptera, but the parietal does not extend on the back of the bar, which is rough there, as in B. musculus, though not so rough. In B. rostrata the backward direction of the bar is still greater than in B. borealis, but the parietal does not reach over any part of it, and the fissure between the two scrolls of the orbital cone is widely open from the root.

and wide palato-maxillary fissure to be noticed with the nasal cavity.[1]

6. MALAR BONE.—In Megaptera the malar bone is much broader (2⅝ inches) along its posterior half than in B. musculus; much more deeply grooved at each end, transversely, to receive the zygoma and maxillary; and its enlargement at the maxillary end (5 inches in breadth, 2½ in length) is abrupt and at right angles to the rest of the bone. In B. musculus the enlargement (4 to 5 inches in length, 3½ in breadth) is more gradual, and is in the direction of the curve of the bone, pushing in below the lachrymal. At the posterior end, in B. musculus, there is an abrupt process, half an inch long, like the end of the little finger, projecting inwards. The outer edge of the malar in B. musculus appears unfinished, as if it had been covered with cartilage.

7. THE ORBIT.—The opening of the orbit is larger in Megaptera, about the same in length, but greater in height (8 inches, in B. musculus 7). This is owing to the much greater curve of the upper edge of the frontal plate in Megaptera (concavity 3 inches deep, and uniform); in B. musculus there is very little curve on the anterior three-fourths of the frontal edge, the bend down being on the posterior fourth. In front of the orbit, the frontal plate presents the same difference for about a third of its breadth, convex in Megaptera, flat in B. musculus. In the roof of the orbit the same difference in the concavity is

---

[1] The parts which assist to close the orbital cone below are more conveniently studied with these parts than with the orbit viewed externally. In B. musculus the two curved laminæ alone form the cone all the way in, the posterior scroll, as in all the finners, below the anterior, where they meet and cross for a little. In Megaptera the cone is covered and as if supported below by the following parts, from within outwards. (1) The suborbital process of the temporal bone, above noted as crossing the pterygoid, supports the pterygoid at the base of the pedicle. (2) The angular process of the palate bone supports the pterygoid from before. (3) The pterygoid bone sends a process outwards below the cone for 3 inches; it rises higher and reaches farther out before than behind, rising to the height of the anterior scroll—that is, an ascent of 1½ inch on the pedicle. (4) The parietal bone, external to the pterygoid, covers the cone below for 3 inches. This is to within 5 inches of where the two scrolls diverge and open the cone outwardly. Over these 3 inches the parietal turns well up in front, to as high as the edge of the anterior scroll. The great extension of the parietal on the back of the cone, the post-orbital bar, has been noted above. The two last-mentioned additions give the orbital cone a blunt keel-shape and smooth surface below, in contrast with its flatness and roughness in B. musculus.

seen, back to the mouth of the narrow orbital cone. The mouth of the cone is ovoid, in Megaptera antero-posteriorly, in B. musculus vertically. The meeting of the two laminæ which close the cone below is, from the orbital margin, in Megaptera 13 inches, in B. musculus 15. The plane of the orbital edge is different. Prolonged forwards the plane would, in Megaptera, cut off the anterior ⅓ or ¼ of the beak; in B. musculus it would clear the beak, passing about a foot external to it. It is evident to the eye that the orbit has much less obliquity forwards in B. musculus than in Megaptera.

8. TRANSVERSE FRONTAL FOSSA.—At the part which may be so named, the supra-occipital, parietals, frontal, nasals, maxillaries, and premaxillaries meet or come in near relation. The anterior edges of the supra-occipital and parietals[1] are seen as two strata lying on the frontal; and, after an interval (the fossa), the frontal meets with the nasals and maxillaries. The gap is 1¾ to 2 inches long in Megaptera, in B. musculus ¾ inch; the depth, ½ to ¾ inch in both. The greater length of the fossa, and its prolongation at the middle, in Megaptera, are mainly owing to the differences at the nasals, to be presently noticed, but the length is the same at the maxillary part as at the top of the nasals. Transversely, the fossa is less in Megaptera than in B. musculus. Here the temporal fossa in Megaptera forms a special inward projection (1 to 1½ inch), gently triangular, as if the skull were pinched at this part, while in B. musculus the inner wall of the fossa goes straight on. The narrowest part between the temporal fossæ is here, at the transverse frontal fossa, in Megaptera 10½ inches, in B. musculus 15½. The great extension of the temporal fossæ inwards in Megaptera, forwards in B. mus-

---

[1] The parietals are not concealed here in any of these skulls. From the usual triangular thickening at the side, where the lateral plate gives off the transverse plate, the latter is seen to pass in to the middle, where, apparently, it is fused with its fellow. In Megaptera the edge of the supra-occipital is about ½ inch thick, but bevelled at the very edge; that of the parietal about ⅛ inch. In B. musculus they are thinner, respectively about ⅙ and ⅛ inch. In B. borealis the edge of the supra-occipital is ¼ to ⅓ inch thick, that of the parietal about ½ inch, and it projects for ⅛ inch or more in front of the supra-occipital. The length of the fossa in B. borealis is 1 inch at the maxillaries, 1½ at the nasals. In B. rostrata, too, the parietal edge is seen all the way across. A difference in Megaptera is, that the supra-occipital and parietal bend down a little as they cross, in accordance with the transverse concavity of the supra-occipital, while in these other finners the edges of the supra-occipital and parietal are raised at the middle.

culus, may in part account for the differences in the nasal bones,
especially for their lateral compression behind in Megaptera.

9. LENGTH AND PROPORTIONS OF THE CRANIUM.—The length
of the cranium, from the back of the occipital condyles, may be
taken to several points. If to the anterior end of the nasals, it is
influenced by the very variable nasal bones; if to the anterior
end of the temporal fossa, it is influenced by the muscular
arrangements. To the posterior end of the nasal bones is probably the best point. These three measurements are given in
the table (Table I. Nos. 45, 46, and 47). Taken at the latter
point the length of the cranium is only $\frac{1}{2}$ inch more in B.
musculus than in Megaptera (32$\frac{1}{2}$ and 32), while in total length
the skull in the B. musculus exceeds that of Megaptera by 20
inches (145 and 125). In breadth the cranium of Megaptera
exceeds that of B. musculus considerably, at the greatest breadth
(measurement No. 15) by 4 to 5 inches (71$\frac{1}{4}$ and 66$\frac{1}{2}$). The
greatest length and greatest breadth of the skull are, respectively in inches, in Megaptera 125 and 71$\frac{1}{4}$, in the B. musculus
145 and 66$\frac{1}{2}$.

10. NASAL BONES.—The nasal bones are very different in Megaptera and B. musculus.[1] The typical form of the nasals in fin-
whales is that which the human nasals would present were they
thickened towards the nasal cavity until the thickness exceeded
the breadth. What would have formed the sixth surface of
a four-sided block is reduced to a border by the sloping of
the superficial and inferior surfaces to an anterior free border.
Differences are seen in the amount of transverse concavity and
compression of the superficial surface, in the concavity of the
free border, in the separation of the inner borders by a trough-like

---

[1] Professor Flower has given (*Proc. Zool. Soc.*, 1864, p. 390) an interesting
series of drawings of the superficial surface of the nasal bones. It is not easy to
give by one view a satisfactory idea of the form of these bones, to show especially
their variously grooved condition. The nasals of this Megaptera differ from his
figure (fig. 3) in that their outer margin, at the anterior end, falls short of the
inner margin at the peak by only 1 inch (about $\frac{1}{4}$ of the whole inner border), and
that the triangular space between the nasals runs all the way, tapering to the
peak where the two nasals come close together. Compared with his figure (fig. 4)
of B. musculus, the outer margins in this B. musculus go much farther forwards,
the inner margins form a prominent narrow mesial peak, and at the posterior
border the obliquity is in the opposite direction, viz., forwards and outwards, at
an angle of 45°.

space or by a process of the frontal, and in the direction of the posterior surface, roofing the fore part of the nasal cavity.

In this *Megaptera* a triangular spine of the frontals, 3 inches long and at first over 2 in breadth, is fitted into the posterior part of the internasal trough, but sunk so as to continue the trough to the top. The visible surface of this frontal spine is furrowed, the furrows on each side running backwards and inwards to the mesial suture. No other of these finners has this nasal spine visible on the surface. The trough between the inner borders runs the whole length, on the posterior half $1\frac{1}{2}$ inch wide, $2\frac{1}{2}$ deep, narrowing towards the point where the nasals meet as a prominent peak. The superficial surface is very narrow behind, the first inch like the end of a large thumb, $1\frac{1}{4}$ inch broad, rounded at the end and sunk into a smoothly arched recess in the frontal. Then the surface broadens by the rapid rise of the inner border, and thereafter becomes grooved and broader distally; breadth at the middle, from edge to edge of the surface, $1\frac{1}{2}$ inch, increased at the hollow of the free border to $3\frac{1}{2}$ inches. On to this point the surface looks more outwards than upwards, the plane of the inner border being about 3 inches higher than that of the outer border. The peak projects fully $2\frac{1}{4}$ inches beyond the hollow of the distal border; its very oblique inferior border meets the straight superior border at an angle of 45°, the point a little rounded off. The inferior or nasal surface is, on its posterior half, nearly flat with a slight slope upwards to its fellow; on the anterior half, it is bevelled and grooved, facing forwards and inwards, forming a low arched roof, and joins the superficial surface at the free end, the junction so rounded off that there is no exact border between the two surfaces. In this Megaptera there is a foramen in the nasals large enough to receive a large goosequill, largest in the right nasal, situated on the anterior surface near the distal border. It is present at about the same place in the right nasal only, in this B. musculus.

[In *B. musculus* the proximal end is transverse for $\frac{3}{4}$ inch close to the middle line, then slopes obliquely outward and forward at an angle of 45°, forming a very jagged articulation with the frontal. This obliquity makes the outer border begin almost 2 inches anterior to the inner. The nasals are at first in close contact and well bevelled up till at about $\frac{1}{2}$ inch from the frontal, the internasal trough begins narrow and shallow (width $\frac{2}{8}$ inch behind, **narrowing** distally, depth $\frac{3}{4}$ inch) compared with that of Megaptera. **External** to the sharp and raised inner border is a shallow **groove**, $\frac{2}{4}$ inch broad, tapering distally. The superficial surface is **broader and more** deeply grooved than in Megaptera; from edge to **edge at the peak,** the depth of **the hollow** is about 2 inches, in Megaptera about **half as much.** The greater depth of the concavity on the distal half in **B. musculus is** owing to the outer margin not being so **low** (only 1 inch **lower than the inner margin** at the peak), and to the surface bending **rapidly down to meet the inferior surface.** The free border of separation **is so far down, and the fall is so rapid, that the distal half of** the superficial surface **looks almost like a distal surface, like the** transversely hollow **base of a four-sided** pyramid. The two nasals together of B. musculus are **not unlike a cocked hat. The peak is as if the distal 2 inches of it in Megaptera had been cut** off vertically; its **two** borders form **just a little less** than **a right** angle, and both **belong to** the superficial surface. A nearly vertical symphysis is thus formed below the peak, about $2\frac{1}{2}$ to 3 inches long, until the sharp distal border is reached which separates the inferior from the superficial surface. The inferior surface is, at the outer part, moderately inclined towards its fellow; on their inner part, rapidly so, making a wedge-shaped recess of the inner part of the roof, 2 inches across, $1\frac{1}{2}$ high.]

These differences between the nasal bones of Megaptera and B. musculus are greater than it is easy to bring out by general measurements or drawings.

11. POSTERIOR NARES.—The form is semilunar, broader at one end, convexity above, the long axis inwards and forwards and a little upwards. The broader end is in Megaptera the inner, in B. musculus the outer. The obliquity is much less in Megaptera, inner end 2 to 3 inches anterior to outer end; in B. musculus 6 inches. They are smaller in Megaptera, long diameter 6 inches; vertically, at outer third 3 inches, at middle $3\frac{1}{2}$, at inner third $3\frac{3}{4}$. In B. musculus the same measurements are respectively $8\frac{1}{2}$, $4\frac{1}{4}$, $4\frac{1}{2}$, 4. The width of the posterior nares together, transversely, opposite their anterior ends, is in Megaptera $12\frac{1}{2}$ inches, in B. musculus $14\frac{1}{4}$ inches.

Parts near the Posterior Nares.—Here there are marked differences. In Megaptera the slanting edge of the *vomer* is shorter (length 8 to 9 inches), and begins close to the posterior end of the bone, as a low ridge, and at the nares it is a 3-inch high septum, and very thin ($\frac{1}{4}$ to $\frac{3}{8}$ inch at mid-height there). In B. musculus, instead of a ridge there is a shallow median groove for 3 or 4 inches, running forwards from a 1½-inch-deep median notch, no ridge proper till the nares are reached, when the vomer presents a posterior edge 2 inches in height (2½ on the slanting edge), thickness at mid-height ¾ inch. Length of vomer from posterior end to septum of nares 14 inches, 1½ more at the side of the notch. The alæ of the vomer in Megaptera go only about a third of the way down on the sides of the great sub-basilar notch, there articulating with the pterygoid by a wavy antero-posterior suture; in B. musculus the alæ go down to within an inch of the lower end of the notch, posteriorly, by a broad triangular plate covering the pterygoid. The sub-basilar notch differs in form in Megaptera, besides being 2 inches broader below, it is broad at the top where it is formed by the vomer, and moderately convex along the sides; in B. musculus it is nearly flat at the top for about 3 inches, and has very little convexity on the sides. The distance of the vomer from the occipital condyle is nearly the same in both (Megaptera 5½ inches, B. musculus 5). The distance from the end of the condyle to the posterior nares, at their septum, shows well the shortness of the cranium here in Megaptera (17½ inches, in B. musculus 24).

The *hamular processes* have been injured in Megaptera, but enough remains to show a marked difference in the broad part. At the notch external to the broad part, the breadth is about the same (Megaptera 4 inches, B. musculus 4½), the difference is in the direction and curvature of the broad part; in Megaptera, directed more inwards and curved inwards, giving a groove on the inner surface into which three fingers may be laid, one behind the other two, while in B. musculus the inner surface is almost flat. This difference accords with the different form of the posterior nares, the outer end of the nares being narrow in Megaptera and broad in B. musculus.

12. ANTERIOR NARIS.—There is considerable difference in the form of the single bony anterior naris; in Megaptera compressed vertically, in B. musculus compressed laterally. In Megaptera the outer wall is bent, the whole wall forming a triangular recess about 2 inches deep, and the roof is but moderately arched. In B. musculus the sides are mostly vertical, rounded below, and the roof is much arched, acutely so towards the middle.

Width of the space (naturally subdivided into the two anterior
nares) in Megaptera 13½ inches, in **B. musculus** 10 inches;
height, from **level of** upper edge of vomer, in Megaptera about
5 inches, in B. musculus averaging about 5½ inches; **height**
from the floor of the vomer, in Megaptera 13 inches, in **B. musculus**
12 inches. That would give the anterior nares in **B. musculus**
as smaller than in Megaptera, although the posterior are larger
than in Megaptera. In **connection** with this, it is to be noted
that the blow-hole **space in front of** the anterior nares is wider
in Megaptera than **in** B. musculus. Also **that the cavity of the**
vomer near the anterior nares is wider as well as deeper in
Megaptera; width in Megaptera, **inner margin,** 7¼ inches, in
B. musculus, at the inverted margin 4¾ **inches, at** the cavity 5½.

13. NASAL CAVITY.—A very marked difference here is in
the much larger deficiency of bone in the **outer wall in B.**
musculus, between the palate, maxillary, and frontal. It is best
seen when viewed from the anterior naris. The fissure in both
goes back to where the frontal forms the anterior scroll of the
orbital cone, and opens **there.** In Megaptera it extends forwards **for 11 inches, of** which **7** are between the palate and
frontal, **the remainder in a recess** in the hinder end of the nasal
plate of the maxillary; height 1 inch, 1½ anteriorly; distance of
anterior end from anterior nares 9 inches. In B. musculus **the**
fissure begins at the anterior **naris, and** as a wide gap, **has a**
length of 17 inches, and a height of **4** inches anteriorly,
diminishing backwards to 3 inches. It is entirely palato-frontal,
except at the anterior end, where it is bounded by **the hinder**
border of the nasal plate of the maxillary. In Megaptera this
plate passes back for 12 inches behind the anterior naris. Seen
from the palatal aspect, this fissure in Megaptera, **7** inches in
length, 1 to 1½ inch in height, is between **the** palate bone
below and the maxillary and frontal above, the latter after the
maxillary terminates at 3 inches **from** the root of the orbital
cone. In B. musculus it is 17 **inches** in length, the upper
boundary **formed by the** palate plate of the maxillary for **13 of**
these, the 4 posterior by the frontal. The total *length of the*
nasal cavity, measured straight from between the upper
margins of the posterior and anterior nares, is **in Megaptera**
30 inches, in B. musculus 23, the shortness in **the** latter owing

mainly to the more anterior position of the posterior nares on the basis cranii, partly to the nasal bones being 1½ inch shorter.

14. ETHMO-TURBINALS.—The lateral mass of the ethmoid bone is more developed in Megaptera, and there are two, if not three, turbinals; in B. musculus only one turbinal, with a rudiment of a second. In Megaptera the lateral mass has a height of 3 to 4 inches, and projects inwards for 1 to 2 inches. The meatuses are—(1) Not seen in a front view, but felt by the finger at the back as a notch and short groove, directed horizontally forwards, rather above the level of the next. (2) Seen on the anterior third of the lateral mass, a groove 2 inches in length, large enough to receive the little finger but wedge-shaped: from below the middle of the mass, directed upwards and forwards, issuing and bifurcating at the anterior pointed end of the mass. Its inner edge is the turbinal, the free edge of the inner convex surface of the mass. (3) An inch below the hinder end of the latter begins the anterior end of the lower meatus, 1 to 1½ inch in width, ½ to ⅓ inch in depth, widening backwards for about 4 inches, where it is lost on the inferior surface of the mass. Its inner overhanging edge is the turbinal, narrow edged, a little curved, with the concavity to the meatus.

In B. musculus there is only the meatus which appears to correspond to the second of those above noted in Megaptera; but it is deeper and fissures the front of the mass so much that the meatus joins the narrow space of the roof. The meatus does not reach quite to the back. The turbinal is a narrow tongue of the lateral mass, with a deep groove between it and the roof of the cavity, which, in both B. musculus and Megaptera, is formed by a curved lamella reaching out from the mesethmoid. In B. musculus the lateral mass is not developed anteriorly and inferiorly, where No. 3 meatus occurs in Megaptera; but at the back the finger detects a notch and short wide groove, which may represent the back part of the 3rd meatus in Megaptera.

The mesethmoid also differs here. In Megaptera it comes down as a thick septum (1½ inch thick above, 3 inches below) nearly vertically for 2 inches and then slopes backwards, with about 1 inch deep excavation; in B. musculus, after a very short median projection, it becomes deeply excavated (3½ inches), the two laminæ 2 to 3 inches apart. The recess below this, which receives the lower part of the back of the mesethmoid cartilage, is, in B. musculus, about 4 inches deep, and large enough to receive the closed hand; in Megaptera it is about 1 inch deep.

15. PRENASAL SPACE.—This wide gap, at and anterior to the blow-holes, is much wider in Megaptera, in which it is a narrow ovoid, in B. musculus elliptical. The width posteriorly, in front of the nasals, is about 10 inches in both; greatest width, in Megaptera 14½ inches, in B. musculus 12½ inches; at the narrow anterior end, in Megaptera 4 inches, in B. musculus 3½. The widest part is, in Megaptera at 13 inches distal to the hollow end of the nasals, 11 inches distal to the temporal fossa, is at the middle of the first quarter of the beak, and just distal to the coronoid process of the premaxillary. In B. musculus the widest part is 17 to 18 inches distal to the hollow end of the nasals, and 6 inches along the beak from the temporal fossa. The widest part is, therefore, some way anterior to the distal end of the blow-holes, supposing these to begin near the hollow anterior border of the nasals. The distal end of the pre-nasal space is better marked off in B. musculus than in Megaptera; length of the space to this part, in Megaptera 30 to 31 inches, which carries it for a third of the way into the second quarter of the beak; length in B. musculus 25 inches, which carries it to about the middle of the first quarter of the beak. The limit is indicated by a rapid contraction to a low rounded angle. It is well to notice here that in these two skulls the width of the narrow inter-premaxillary space at the middle of the beak is, in Megaptera, 3 inches, in B. musculus 1¼.

The *vomer* here, the widest part of the space, is a little deeper in Megaptera, but differs considerably in width and form; contracting at the upper edge in B. musculus to 5¾ from 6⅝ some way below, while in Megaptera the upper edge is scarcely inverted, and the width is 7⅜ inches, in adaptation to the greater width of the prenasal space in Megaptera. The upper edge of the vomer also differs in thickness. In Megaptera it is thickest (1½ inch) at the back part of the prenasal space, and continues thus thick for some way within the nasal fossa; in B. musculus it is thickest (2¼ inches) at the middle of the prenasal space, and diminishes in thickness forwards and backwards.

16. MAXILLARY AND PREMAXILLARY BONES AT THE NASAL REGION.—While in B. musculus the profile of these bones is here almost straight on to the beak, in Megaptera they present a marked fall from their posterior end for 7 or 8 inches along the beak, and from this hollow the *premaxillary* sends up a process which may be termed its *coronoid process*. Height of process 1½ inch, base about 12 inches in length, beginning 2 inches anterior to the peak of the nasal bones; anterior slope of process the longest, summit from 5 to 6 inches anterior to the nasal peak. The height to which this process of the premaxillary rises in the hollow is just to the level of a straight line drawn from the top of the frontal process to the beak. The elevation of the surface of the head, referred to with the external characters, would appear to be a little anterior to this long elevation, as the blow-holes, 11 inches in length, were situated on the hinder slope of the elevation.

In Megaptera the parts of the premaxillary and maxillary at the side of the nasals and anterior to them, as they pass backwards are much inclined inwards, and bent with the convexity inwards, while in B. musculus they have very little inclination inwards. A line prolonged in the direction of the inner edge of the maxillary, from the beak, goes external to the frontal end of the frontal process—in Megaptera 5 inches, in B. musculus 1 inch. The contraction from the widest part of the prenasal space to between the frontal ends of the maxillaries is, in Megaptera from 14½ inches at the former place to 4 inches at the latter; in B. musculus from 12½ inches to 5. This, in Megaptera, is partly owing to the greater width of the prenasal space, partly to the greater inward encroachment of the temporal fossa at this part. Width between the frontal ends of the maxillaries, outer border, in Megaptera 11 inches, in B. musculus 14½. The frontal process of the maxillary is broader in Megaptera, and increases in breadth distally; at the top 3¼ inches, at 6 inches forwards 4 inches; and then sweeps broadly outwards in its distal half as the edge of the temporal fossa.

[In B. musculus the breadth of the frontal process of the maxillary at the top is 4 inches, at 12 inches forward only 2¼ inches, owing to the forward extension of the temporal fossa. The surface of the process is very convex transversely in Megaptera; in B. musculus it has very little transverse convexity.]

17. Ant-Orbital Process of the Maxillary.—The great length and backward and downward sweep of this process from the beak in Megaptera is characteristic. Length of process from beak 18 inches, with a fall of about 12 inches; in B. musculus length about 15 inches, less as seen in front, with a fall of about 6 inches. The form also is different, prismatic in Megaptera, flat in B. musculus. In B. musculus, where the process goes off, the curved border of the temporal fossa rises into a tubercle 2 inches high, 3 inches broad at the base. The border then twists forwards soon, overlapping the border from the beak, and forms the anterior edge of the flat process. This is the result of the forward extension of the temporal fossa in B. musculus. In Megaptera the prismatic form is owing to a lamina rising upwards and backwards, the edge of which is continued from the front edge of the temporal fossa. This lamina is represented in B. musculus by a low smooth ridge on the flat temporal fossa surface of the maxillary at 3 inches behind the anterior border of the process, while the tubercle in B. musculus is represented in Megaptera by a sharp even ridge running along the facial surface of the process. The outer end of the process in Megaptera is, in length externally, in front of the malar, 3 inches; in breadth, within the orbit, 6 inches; in B. musculus length 4½ inches, breadth within the orbit 6 inches.

The *overlapping of the frontal by the maxillary* is greater in Megaptera; breadth of the part overlapped 7 to 8 inches, bevelled marking on frontal 3 inches in length, but only 2 inches of it covered by the maxillary. In B. musculus breadth of angle overlapped 6 inches, the length overlapped 2 inches; no bevelled marking on frontal. Length of maxillary uncovered by frontal in temporal fossa, in Megaptera 1 inch, in B. musculus 4 inches. The ant-orbital process is in Megaptera 1 to 1½ inch behind the frontal process; in B. musculus 9 inches in front of it.

18. The Beak.—A survey of the beak from before at once shows the well-known greater breadth, with more convex edges, in Megaptera, contrasting with its long tapering form in B. musculus. The actual length in Megaptera is 85 inches, in B. musculus 96. It is convenient to divide the beak into quarters. At the base of the beak the breadth, along the curve, is less in

Megaptera than in B. musculus (45 and 48 inches), and at the end of the first quarter (Megaptera 33½, B. musculus 36); but after this the breadth is greatest in Megaptera, at the middle Megaptera 27½, B. musculus 27; at the end of the third quarter, Megaptera 19¼, B. musculus 17. The measurements given in the table show to what extent these breadths belong to the maxillary, premaxillary, or to the mesial space. At the base, and at the end of the first quarter, the prenasal space and the direction of the surface of the premaxillary affect the measurements. At the middle of the beak the greater breadth in Megaptera is owing to the mesial space, which is nearly 2 inches wider than in B. musculus, but at the end of the third quarter nearly half of the decidedly greater breadth in Megaptera than in B. musculus is owing to the maxillary. Viewed from above, the breadth of the beak in Megaptera has very much the appearance of being less in front of the base than at the end of the first fourth, but the form deceives the eye; the narrowest part is behind the middle of the first quarter (33 inches), and from that part the breadth diminishes forwards.

The fall of the upper surface of the beak to its outer edge is on the posterior half, greatest in B. musculus, on the anterior half greatest in Megaptera. The amount of the fall at the base, at the end of the first quarter, at the middle, and at the end of the third quarter, respectively, is, in inches, in Megaptera 10, 4½, 3⅜, 2½; in B. musculus 10½, 5, 3, 1⅞. This greater slope of the distal half of the beak in Megaptera is manifest to the eye. The surface is less convex transversely in B. musculus, giving the distal half of the beak a very flat appearance in B. musculus. In Megaptera the maxillary is convex transversely about the middle, with a slight concavity internal to this, but the chief transverse convexity is on the premaxillary, which is so great, on the second quarter and on part of the third quarter, that the highest part of the beak is on the surface, not at the inner border, as it is along the distal half of the beak. The line of articulation between the maxillary and premaxillary is, in Megaptera concave outwards on the two middle quarters (concavity 1¼ inch deep), convex outwards on the distal quarter. In B. musculus these undulations are much more gentle, the concavity of the first about ¾ inch deep.

Premaxillary Bone.—Besides the transverse convexity above noted, the premaxillary in Megaptera is later in undergoing the seeming twist of the surface. In B. musculus the premaxillary, very concave along the distal half of the prenasal space, becomes nearly horizontal at about a third of the distance into the second quarter of the beak, and also nearly flat, except that there is a shallow groove along its inner third (prolonged from the concavity at the prenasal space) and a gentle convexity on the outer third.

In Megaptera the premaxillary on the distal half of the prenasal space is scarcely concave, and at the end of the first quarter of the beak the surface looks more inwards than upwards, and does not become horizontal till near the middle of the beak.

This seeming twist of the premaxillary, however, is not in reality a torsion of a flat bone, but is owing to the development of a transverse plate, beginning on the distal half of the prenasal space, which goes inwards and contracts that space. The premaxillary thus attains a sharply triangular form in transverse section. The internal or nasal surface is very concave in Megaptera from the end of the prenasal space onwards; in B. musculus the cavity does not begin till near the middle of the beak. In Megaptera the internal lamina is much more developed on the second quarter of the beak than onwards from this, the approximation of the premaxillaries being accomplished along the distal half by the more inward position and more inward slope of the vertical plate of the bone. In B. musculus the inward plate of the premaxillary appears to overhang to about the same extent throughout.

The *width of the inter-premaxillary space* along the distal three-fourths of the beak is, at the beginning of each quarter, respectively, in inches, in Megaptera, $8\frac{1}{2}$, 3, $2\frac{1}{2}$; in B. Musculus, $2\frac{1}{4}$, $1\frac{1}{8}$, $1\frac{3}{4}$. On the distal quarter, in Megaptera it continues to contract gradually to $1\frac{1}{4}$ inch at the point; in B. musculus it at first widens to 2 inches, and then contracts, width at the point about $\frac{3}{4}$ inch.

19. FORAMINA ON THE FACIAL SURFACE OF THE MAXILLARY. —These large foramina present differences in Megaptera and B. musculus, but reliance cannot be placed on characters which

differ considerably on the two sides of the same skull. In Megaptera, however, they are larger, extend more into the second quarter of the beak, and are on the whole more internal in position than in B. musculus. In the latter they are 9 in number on each side, not symmetrically placed, and are on the first quarter of the beak, 5 inches within the line, except one on the left side, which is on the second quarter 3 inches beyond the line. In Megaptera they are, on the right side, 7 large, 3 very small; on the left side, 10 large and one very small. Even at the base of the frontal process they are not symmetrical; right side, one large foramen (corresponding to the infra-orbital foramen in man), directed backwards and outwards, admitting two fingers; on left side, three foramina, each admitting a finger, the upper one with the reverted direction. The other foramina are too unsymmetrical to admit of individual comparison. On the left side two are on the line between the first and second quarters of the beak, and two are respectively 3 and 5 inches into the second quarter; on the right side one is on the line and one 3 inches beyond it on the second quarter. The most distal one is large on the left side, three times as large as on the right side. The foramina on the palatal aspect are more distinctive.

20. THE BEAK ON THE PALATAL ASPECT.—The chief differences here are the greater breadth of the median beam and the partial absence of the vascular grooves in Megaptera. The measurements given in the table show also that the depth of the hollow on each side of the median beam is much less on the posterior part of the beak in Megaptera. The narrow keel-like median beam in B. musculus gives it the appearance of greater projection, but the following measurements, taken at the junction of the first and second quarters of the beak, will show that the projection is less, and also the greater thickness of the beam in Megaptera. Median beam below level of outer edge of beak, in Megaptera $10\frac{1}{2}$ inches, in B. musculus $8\frac{1}{4}$; breadths of median beam, at two inches up, Megaptera $3\frac{3}{4}$ inches, B. musculus 3; at mid-height, 6 inches up on the slope, Megaptera 8 inches, B. musculus 6; at level of outer margin of beak (12 inches up, on the slope, in Megaptera, in B. musculus $8\frac{1}{2}$ inches up), Megaptera 15 inches, B. musculus $8\frac{1}{4}$; concavity taken on

level with outer margin of beak, in Megaptera, width 8 inches, depth 1; in B. musculus, width 12¾ inches, depth 2.

Seen in profile, the lower edge of the median beam of the palate is less curved in Megaptera; depth of concavity in **Megaptera** 3 inches, along the maxillary part only, 2 inches; depth in B. musculus 4 inches, along the maxillary part only, 2 **inches**.

The breadth of vomer exposed between the lower **edges** **of the** maxillaries is, in Megaptera about 1½ inch all along, except about the middle when it is increased to 2¼ inches behind, between the palatals, ½ inch in vomer is seen, increasing **forwards.** In B. **musculus, along the** anterior half, about ½ inch, **on the** posterior half, ¾ **to 1 inch in breadth of** the vomer is **seen.** The bony vomer extends to within 27 inches of the point of the beak in Megaptera, to within 15 inches in B. musculus.

21. VASCULAR GROOVES ON THE PALATÀL SURFACE **OF THE** MAXILLARY BONE.—This system of great palatal grooves, and their foramina, will be better understood by observing them first in *B. musculus.* They are great grooves ⅔ to ½ inch broad, such **as** might be made with the end of the finger on a soft **surface. They may be** classified as (*a*) those of the roof, belonging to the whalebone region, and (*b*) those of the median beam. **Those of the latter (*b*) issue** from the fore-end of the palato-**maxillary fissure, at least 3** in number, descend along the beam **with more or less obliquity,** one of the three arched up and **covered for** a time **on the** right side, on the left side **two are** **thus arched** and covered. The most anterior one reaches the lower edge of the beam at about the middle of the second quarter of the beak, and ceases at the middle **of the beak.** (*a*) Those of the whalebone region may be classified as anterior or longitudinal, seen on more than the anterior three-fourths of the **beak; the** intermediate, seen on the posterior half of the first quarter of the beak; and the posterior, seen below the temporal **fossa.** Of the *anterior or longitudinal series,* three issue **on the** anterior half of the first quarter of the beak, the external first, **the** third at the end of the first quarter, a fourth **some** way along **the second quarter.** The two last of **these run** **along** the inner part of the roof, where the roof meets **the median beam,** close together **as** great grooves, half an **inch in** breadth **and deep** enough to receive half the thickness of the finger, the

one that is last to appear going on to the point of the maxillary, seen along a course of about 5 feet. The *intermediate series*, 3 in number, issue in the roof 9 to 8 inches from the outer edge, pass obliquely outwards and forwards for about 6 inches, bifurcate and cease at about 3 to 4 inches from the outer edge, at the first quarter of the beak. The *posterior series*, 3 or 4 in number, issue in a line with the latter series, below the temporal fossa, back to near the hinder edge of the maxillary plate. They pass forwards and outwards, more curved than the intermediate series (concavity backwards), bifurcate, and may then curve partly backwards; but in B. musculus (in contrast with the posterior series in B. borealis) these sub-temporal grooves do not at first turn backwards, although at the very back, where the bone is much perforated and scaly, there may be a small exception to this. They, too, have a course of about 6 inches before they bifurcate and cease.

In *Megaptera* this grand system of grooves is deficient except the longitudinal series along the beak, and partly on the median beam, being in other parts replaced by foramina only with occasional short grooves. On the median beam two grooves are seen to begin at the fore-end of the palato-maxillary fissure, the lower one, the greatest and longest, 1 inch broad, passing downwards, and lost before the end of the first quarter of the beak is reached. Three longitudinal roof-grooves appear successively on the second quarter of the beak, and one runs on to the end. The intermediate roof-series are represented on the first quarter of the beak by apertures with very short grooves, one large aperture and two or three smaller, at distances of only 4, 5, and 6 inches from the outer edge of the beak. In the sub-temporal region, there are the foramina and scales of bones overhanging shallow spaces, but not a system of grooves like that of B. musculus. The cause of this deficiency of intermediate and posterior roof-grooves in Megaptera is the great breadth of its median beam, narrowing and filling· up the concavity on each side, and more or less roofing over what are grooves in B. musculus. The position of the apertures of the intermediate and posterior series, it will have been noticed, is, accordingly, much farther out in Megaptera than in B. musculus.

22. CRANIAL CAVITY.—The measurements given in the table

show that the height is the same in both, that in Megaptera the breadth is greater by ⅓ part (2 inches), the length less by ¼ part (4 inches) than in B. musculus. This accords with the proportions of the back part of the skull externally, but the whole cavity appears to be more capacious in B. musculus. The opening of the *olfactory fossa* is triangular in form, base below, and smaller in Megaptera; in B. musculus bluntly triangular, base above with a narrow notch at the middle (Megaptera, breadth 3 inches, height 2; B. musculus, breadth 3¼, height 3). The fossa itself in Megaptera is directed more upwards, is curved, and is shorter than in B. musculus (Megaptera 4 inches, B. musculus 5).

Other Characters within the Cranium.—In Megaptera the suture between the post and pre-sphenoid (10 inches from the foramen magnum) is open across its whole breadth; in B. musculus there is a short transverse ridge at the middle in the corresponding position (12 inches from foramen magnum), but no suture visible. Where the basi-occipital and post-sphenoid appear to have united, there is in Megaptera (5 inches from the foramen magnum) a curved ridge, as prominent nearly as a finger laid on flat, curved, convexity backwards; in front of it a wide shallow fossa; going back from it a similarly raised median ridge, on each side of which is a rounded fossa. In B. musculus the transverse ridge (8 inches from the foramen magnum) is very low, the median ridge behind is well marked but broad, and there is no fossa at the side of it. The sella turcica, about 3 inches long in both, is better marked in Megaptera, having a transverse depression on its anterior half; in B. musculus there is rather a transverse convexity, with a slight longitudinal median depression. The common orbital foramen, representing the optic foramen and sphenoidal fissure (which are separate in B. rostrata), is in Megaptera partly divided into optic and sphenoidal fissure parts by a well-marked peak of bone above and below, the interspheniod suture intersecting the lower peak. If this partial subdivision did not exist in Megaptera, the common foramen would form a triangle wider below than the common foramen in B. musculus. In the roof of the cranial cavity there is a well-marked sharp median ridge in B. musculus, much less developed in Megaptera.

PART IV.—*continued.*

Ear Bones, Mandible, and Hyoid.

23. Ear Bone—*Tympanic.***—**The following are the dimensions of the tympanic bone in Megaptera, and in the 50-feet·long B. musculus, in inches :—

	Megaptera.	B. musculus.
1. Length,	$4\frac{3}{8}$	$4\frac{2}{8}$
2. Breadth, on posterior third,	$2\frac{1}{2}$	$2\frac{1}{4}$
3. ,, on anterior third,	2	$2\frac{3}{8}$
4. Height, from tip of lamella before meatus,	$3\frac{1}{2}$	$3\frac{1}{4}$
5. Breadth of smooth part of inner surface, at middle,	$1\frac{5}{8}$	$1\frac{1}{2}$
6. Width of anterior division of the aperture,	1	$\frac{7}{8}$
7. ,, of aperture at the division,	$\frac{7}{16}$	$\frac{7}{16}$
8. Weight, in ounces,	$14\frac{1}{2}$	$15\frac{3}{4}$

The tympanic bone of Megaptera may be distinguished from that of B. musculus by the following characters :—1. It is shorter in proportion to its breadth than in B. musculus. 2. As seen from below, in B. musculus the breadth of the bone at the lobe in front of the median constriction on the outer side is nearly as great as at the lobe behind it. In Megaptera the constriction is farther forward, and the breadth is considerably less before than behind the constriction. 3. Viewed from behind, the surfaces in B. musculus are seen to slope from the inferior border like a gable-roof. This flattening is wanting in Megaptera, the whole of the under surface being convex transversely, except slightly before and behind. 4. Seen on the inner surface, in B. musculus the smooth part is marked off from the rough part by a much more defined border than in Megaptera. 5. This smooth part is more bulged in Megaptera on its posterior half, contributing to the increased breadth on the posterior half of the bone. 6. The aperture is more constricted in Megaptera at the division into anterior and posterior parts. 7. The anterior (Eustachian) division presents marked differences in form. In B. musculus its anterior end bends upwards to a subacute point, so that the lower boundary of the aperture is nearly uniformly concave. In Megaptera the anterior end is directed obliquely downwards and forwards, and is almost square-shaped, so that the lower boundary is convex behind, and then very concave in front. The width of the anterior division of the opening is greater in Megaptera than in B. musculus; greatest width at its posterior half, owing to a bend up of the wall here at the anterior junction with the petrosal, much more marked than in B. musculus. The ridge which here forms the exact upper boundary of this division of the aperture is smooth and concave in B. musculus. It is very thin and sharp in Megaptera, and goes spirally forwards on the smooth surface in front. 8. The width of the posterior (meatus externus) part of the aperture is also greater than in B. musculus, at its last half inch $\frac{1}{8}$ to $\frac{1}{6}$ inch in B. musculus, $\frac{3}{8}$ to $\frac{1}{2}$ inch in Megaptera.

Viewed from the outer side, two depressions are seen, dividing the surface on its upper half into three lobes. The anterior depression corresponds to the constriction between the Eustachian and the meatus divisions of the aperture. The middle lobe is continuous with the broadest part of the bone, and, upwards, there is prolonged from it, at its back part, a flat lamella which projects in front of the external meatus. The third or posterior lobe does not extend so far on the under surface as the other lobes, and is continued upwards to end in a nipple-shaped process, which forms the projecting lower boundary of the meatus externus. 9. In B. musculus the pre-meatus lamella is more bent, sigmoid behind (concavity below the middle). 10. An antero-posterior depression in front of the base of the lamella, bounding the middle lobe above, is much better marked in B. musculus, cutting the lamella off from the lobe, except behind.

11. The nipple-like process projects more in B. **musculus**, projecting for $\frac{2}{3}$ the length of the lamella; in Megaptera projecting for about $\frac{1}{3}$ the length of the lamella. The depression between the nipple-like process and the lamella is narrower in B. **musculus**. 12. The **form of the bone behind the nipple-like process is** different. There is more of the bone behind the process in B. musculus; but what is most distinctive here is, that the inferior border in B. musculus is carried farther round, to the level of the base of the nipple-like process, and as a thick crest; while in Megaptera the border forms a sharp peak at about the middle of the posterior end of the bone.

Periotic.—In Megaptera, the dimensions of the periotic bone are, in inches—*anterior division*, length $4\frac{1}{2}$, transversely 4, height $2\frac{1}{4}$—*posterior division*, length $7\frac{3}{4}$; height $2\frac{1}{4}$ anteriorly, 5 at the middle; thickness $1\frac{1}{2}$ above, below about $\frac{1}{2}$ inch. The **inner surface of the wing-like posterior division is very rough on** its upper half; on its lower **half it is concave vertically and** comparatively smooth, but **is again roughened and thickened at the lower free border.** The whole external surface **is strongly** streaked in the direction of the wing, the ridges curved, convexity downwards. The whole wing is slightly bent on its axis, concavity on the outer surface. Upper border a little concave for $4\frac{1}{2}$ inches (then broken for 3 inches), lower border very convex for 6 inches (then broken off). There is no trace of separate development **of the upper and lower parts of** the wing.[1] The posterior junction between the periotic and tympanic is 1 inch in breadth (outwards and backwards), in thickness $\frac{3}{8}$, thinner behind; the anterior junction is 1 inch antero-posteriorly, in thickness $\frac{1}{4}$ inch in front, $\frac{1}{12}$ inch behind.

Meatus externus and Tympanum.—In Megaptera, the irregularly square-shaped **meatus** externus, included between the tympanic and periotic and their two junctions, measures, at its **outer end**, vertically **1 to 1$\frac{1}{4}$** inch; antero-posteriorly, at the **lower part** $\frac{7}{8}$ inch, below the middle nearly $\frac{1}{8}$ inch less, owing to the backward curve of the lamella. The upper boundary, formed by the periotic, **is** a little concave; the posterior boundary, formed by the posterior junction, **is a little** concave and is smooth; the lower boundary, formed by the

[1] See note, page 187.

tympanic, has projecting into it, for about $\frac{1}{4}$ inch, the nipple-like process. The anterior boundary is formed at its upper $\frac{2}{3}$ by the concave outer edge of the lamella of the tympanic. At the upper $\frac{1}{3}$ of this boundary the bony meatus opens into a fissure, $\frac{3}{4}$ inch long, $\frac{1}{4}$ inch wide, between the periotic and tympanic, closed in front by the hinder edge of the anterior junction, near to which the fissure becomes suddenly narrowed.

The lower bony wall of the *passage*, from its outer end to the posterior division of the opening of the tympanic bone, is short, $\frac{1}{3}$ inch behind, at the nipple-shaped process; in front of this it is a mere edge. The smooth concave surface on the front of the posterior junction is 1 inch in height as well as in breadth, and has that length as a wall of the meatus in to the very back of the opening of the tympanic bone, but about half that smooth surface is external to the plane of the upper boundary. The distances from the outer end of the meatus to the promontory, just below the fenestra ovalis, are at anterior and upper walls of meatus $1\frac{1}{2}$ inch, at posterior wall $1\frac{3}{4}$ inch, taken from the outer edge of the smooth surface. The fenestra ovalis is opposite the upper posterior corner of the meatus externus. It lies at the bottom of a conical recess, about $\frac{1}{3}$ inch deep, into which the stapes is sunk. The stapes moves freely, but it is so held in the fenestra, at the apex of the fossa, that it cannot be removed.

[In *B. musculus* the smooth anterior surface of the posterior junction, bounding the meatus behind, presents a very different character. It forms a deep rounded fossa, about $\frac{3}{4}$ inch in diameter and $\frac{1}{2}$ inch deep, receiving the end of the thumb, and opens in front into the narrow posterior end of the opening of the tympanic bone. In Megaptera the concavity is shallow, about $\frac{1}{2}$ inch deep.]

24. THE MANDIBLE.—The following table (Table II.) brings out various differences on the mandible of the two species. In comparing the form, the parts to be observed are the coronoid process, the condyle, the elongated neck between these two processes, and the body in front of the coronoid process.

TABLE II. *Measurements of the Mandible,* given in inches.

	Megaptera, 40 feet long.	B. musculus, 50 feet long.
*1. Length, in a straight line,	$120\frac{3}{4}$	135
2. ,, along curve, on outer side,	$130\frac{3}{4}$	143
*3. Depth of curve,	$17\frac{3}{4}$	$17\frac{3}{4}$
4. Height, at 6 inches from symphysis,	$6\frac{1}{2}$	7
*5. ,, at middle,	10	9
6. ,, at 12 inches from tip of coronoid,	11	12
*7. ,, at coronoid (body and process),	$12\frac{3}{4}$	17
8. Coronoid process, height behind,	$3\frac{1}{4}$	$8\frac{1}{2}$
9. ,, height in front,	$1\frac{1}{2}$	$4\frac{1}{4}$
10. ,, breadth at base,	7	$8\frac{1}{2}$
11. ,, ,, at mid-height,	$3\frac{3}{4}$	5
12. ,, thickness at middle,	$1\frac{1}{4}$	$1\frac{7}{8}$
13. From top of coronoid to top of condyle,	22	22
14. ,, to end of condyle,	25	25
15. Greatest depth of neck,	$2\frac{3}{4}$	$7\frac{1}{2}$
16. End of condyle to anterior edge of dental foramen,	$11\frac{3}{4}$	16
17. From same edge of foramen to back of coronoid,	9	4
18. Height at anterior fourth of neck,	$9\frac{3}{4}$	$9\frac{1}{4}$
19. ,, at middle of neck,	$9\frac{1}{2}$	$8\frac{3}{4}$
20. ,, at posterior fourth of neck,	$7\frac{3}{4}$	$9\frac{1}{2}$
21. Condyle, height (entire end of bone),	11	$12\frac{1}{2}$
22. ,, thickness, at middle,	6	7
23. Thickness of mandible, near symphysis,	4	$3\frac{1}{2}$
24. ,, at middle,	5	$5\frac{1}{4}$
25. ,, below coronoid process,	$5\frac{1}{2}$	$5\frac{1}{2}$
26. ,, at middle of neck,	5	$4\frac{1}{2}$
For Relation of Mandible to Skull.		
27. From middle of glenoid fossa to end of beak,	117	137
28. From lower part of ditto to ditto,	$120\frac{1}{2}$	140
29. Middle of glenoid fossa to vertical plane of anterior edge of temporal passage,	8	14
30. Ditto to anterior border of floor of orbit,	16	27
31. Projection of mandible beyond beak when condyle is in contact with middle of glenoid fossa,	$2\frac{1}{4}$	2 short

25. CHARACTERS OF THE MANDIBLE IN COMPARISON WITH THOSE OF B. MUSCULUS.

Coronoid Process.—[In *B. musculus* the coronoid process is high ($8\frac{1}{2}$ inches behind, $4\frac{1}{4}$ in front) and curved forwards. The anterior margin has a concavity of about 1 inch deep; the posterior margin is very convex on its upper half, concave (concavity $\frac{1}{2}$ inch deep) on its lower half. The inner surface is nearly flat; the outer surface is concave vertically. The process is moderately everted on its upper half, but mainly owing to the concavity of its outer surface. The outer surface has a strong beam, marking off the anterior third as a deep groove on the upper third of the process, large enough to receive a large thumb. By means of this beam the process retains its thickness till close to the top. The beam goes nearly straight down, and is lost at the base of the process. The inner edge of the groove is the anterior border of the process, and is continued as the anterior border of the body of the bone. The border of the process is sharp. Just in front of the base of the process the border of the bone becomes rapidly thickened. This *pre-coronoid thickening* extends for about 5 inches, and is then gradually continued on the upper edge of the body. Viewed from the side this part is slightly convex upwards. The posterior edge of the coronoid process is thin at the middle third, and on about the lower fourth becomes thickened to $\frac{3}{4}$ inch, as the anterior part of the post-coronoid roughness. The rounding of the top of the process corresponds pretty well to the form of the ends of the four fingers laid together, the fore finger to the front. Viewed from above, the top has a thick, somewhat triangular form, owing to the beam and the groove.]

In Megaptera, besides its much less height than in B. musculus (in Megaptera $3\frac{1}{4}$ inches behind, $1\frac{1}{2}$ in front), the coronoid process differs in form from that of B. musculus. It is a blunt triangle, the anterior and posterior borders both gently concave till near the top, the anterior the most sloping border, so that the process seems to point a little backwards. It is more everted than in B. musculus, and a little concave on the outer surface. The posterior border of the process is thin throughout ($\frac{3}{4}$ inch, increasing to 1 inch below). The top is less pointed than in B. musculus, and is thick, averaging $\frac{3}{4}$ inch, more behind, thinner in front. There is no beam on the outer surface, except a little thickening at the top, and consequently no groove on this surface. The *pre-coronoid thickening* is more marked than in B. musculus, 8 to 9 inches in length, in its posterior half 1 inch thick, and raised $\frac{1}{2}$ inch above the level of the border behind and before. It is rough, with an irregular depression dividing it longitudinally, the outer edge of the depression continuous with the anterior edge of the coronoid process. The depression runs back on the inner side of the anterior border of the coronoid, leaving a groove between, which, however, does not correspond to the groove noted in B. musculus.

Neck.—The neck is about the same length in both, but an elevation on the upper border in Megaptera, at and in front of the middle, forms a marked difference, in contrast with the general gentle concavity in B. musculus. This *post-coronoid elevation* is about 7 to 8 inches in length, rising gradually before and behind to a height of 1 inch above the level of the rest of the upper border of the neck. Viewed from above, it is about $2\frac{1}{2}$ inches thick at the top, narrowing forwards, with a smooth interval, for a hand's breadth, between the elevation and the coronoid process. Backwards the rough surface is suddenly narrowed by the notching to form the dental foramen. The roughness extends down on the inside for 8 inches, in front of the dental foramen, most marked on its anterior 4 inches, where the wall rises as a low mound. The roughness of this post-coronoid elevation is not that of cartilage-covered bone, but as of bone to which tendon or ligament had been attached. The same of the pre-coronoid thickening, but the roughness is less marked.

[In *B. musculus* there is no post-coronoid upward elevation, but on the inside, at this part of the neck, the inward elevation and roughness, over the dental canal, is more marked than in Megaptera. It goes down, for $2\frac{1}{2}$ inches, only to below the level of the middle of the spine, and ends by a well-defined edge. The upper border of the neck at this part, though not forming an upward elevation, is flattened and rough (with a breadth of 1 inch) for 6 inches, being the space between the dental foramen and the coronoid process, the flattening reaching for 2 inches up on the hinder border of the coronoid process. This flattening and roughness of the border here contrasts strongly with the smoothness of this part of the border in Megaptera.]

Direction of the Neck and Position of the Dental Foramen. —The place where the axis of the body and the axis of the neck meet is farther back in Megaptera than in B. musculus. The distance between the back of the condyle and the coronoid process is the same in both (25 inches), but the dental foramen is much farther back in Megaptera ($11\frac{3}{4}$ inches from the condyle) than in B. musculus (16 inches). The axes of the body and neck meet at the dental foramen. This gives a very different form to the necks. The outward convexity of the body in B. musculus ceases just behind the coronoid process, and sharply, so that the outer surface of the neck is deeply concave longitudinally (concavity between the coronoid part and the tuberosity below the condyle, 3 inches deep); while in Megaptera the outward convexity of the body goes back to about the middle of the neck, and the concavity is less (about 2 inches deep).

There is also much greater torsion of the neck in B. mus-

culus. In Megaptera the outer surface is directed a little
upwards, and is gently and nearly uniformly convex. In B.
musculus the upper part of the neck is inclined inwards, the
lower part outwards. The whole outer surface of the neck is
thus more directed upwards than in Megaptera. Also, in B.
musculus the upper half of the posterior third of the neck is
concave, and the lower third of the anterior half is flat. This
torsion of the neck in B. musculus, mainly from the inward inclination of its upper part and of the base of the coronoid process,
is a marked distinctive character. In Megaptera the outer
surface of the neck, vertically, remains little different from that
of the body till close to the condyle.

Viewed on the inner side, the same difference in the vertical
direction is seen, and the axes of the body and neck are seen to
meet abruptly at the anterior edge of the dental foramen in B. musculus, while in Megaptera the change is gradual and not great.

Differences at the Dental Foramen.—(1) It is much nearer
the condyle in Megaptera; from the back of the condyle, 11¾
inches in Megaptera, in B. musculus 16. (2) As seen from
behind, it is on a plane more internal in its relation to the
coronoid process in Megaptera. A vertical straightforward
plane passing through the outer side of the foramen, in Megaptera passes to the inner side of the coronoid process, in B.
musculus intersects the process. (3) The spine, prolonged
back from the inner edge of the foramen, is short (⅝ inch in
length) in Megaptera, in B. musculus long (2¾ inch) and
pointed. (4) The notch below the spine goes forward beyond
the upper notch, in Megaptera ¼ to ½ inch, in B. musculus 2
inches. (5) The groove (about 4 inches long) which ends in
the foramen has in Megaptera a broad concave floor, 2 inches
broad at the middle, the inner edge raised; in B. musculus it
is only 1 inch broad at the middle, and slopes downwards.
The wall above the groove in Megaptera slopes upwards and
outwards; in B. musculus it is vertical or overhanging. The
result is that the groove appears to the eye as if dug out from
above in Megaptera and as if dug out from the inside in B.
musculus. (6) The foramen is larger in Megaptera, vertically
3 inches, transversely 2½; in B. musculus it is ½ inch less in
both directions. (7) What appears to be the mylo-hyoid groove

is present in B. musculus, running on from the lower notch, for 2 inches as a shallow triangle ; it is absent in Megaptera.

Condyle.—The condyle proper is to be distinguished from the more projecting tuberosity below it. The differences on the condyle are—(1) The obliquity, downwards and outwards, is much greater in B. **musculus,** in adaptation to the corresponding obliquity of the neck, above-noted. **(2) In B.** musculus the outer edge of the condyle projects forwards, at its lower half, as a broad thick tongue, overhanging the surface of the neck, receiving the finger between them. In Megaptera, although the outer side of the condyle projects more than in **B.** musculus, there is no abrupt or forward projection, only the thick outer side of the condyle projecting uniformly. (3) The condyle on the inner side in B. musculus projects uniformly, without sharp edge, in its whole height here (about 7 inches). In Megaptera it projects less, has a rough sharp edge at the middle, and is short vertically (about $5\frac{1}{2}$ inches). (4) The groove between the condyle and tuberosity, on the inner side, is in Megaptera situated at the mid-height of the end of the bone, is well defined, large enough to receive the end of the thumb, and is directed obliquely upwards and forwards. In B. musculus it is a wide valley, 4 inches wide, $\frac{3}{4}$ inch deep at the deepest part, which is some way below the middle of the entire end of the bone. (5) The separating groove on the outer side is also more definitely marked in Megaptera, both above, by the more abrupt projection of the condyle, and below, by the more outward position of this part of the tuberosity.

The differences on the *tuberosity* are not very definite, except in regard to the separating grooves above noted ; but it projects more laterally in Megaptera than in B. musculus, both internally, as a sharp forward-ascending edge (entirely wanting in **B.** musculus), and on the outer side as a broadly-projecting outer edge to the bone below the condyle. In B. musculus it is wanting below the inner third of the condyle ; in Megaptera it extends across the whole breadth of the condyle, at the inner part as a narrow ascending edge.

Body.—The proportions of the body of the mandible in Megaptera and B. musculus are seen in the table. The actual *depth of the curve* is nearly the same in both, but as this depth

(17⅜ inches in Megaptera, 17⅝ in B. musculus) is obtained on a
shorter jaw in Megaptera (120¾ in Megaptera, 135 in B. musculus), the curving is greater in Megaptera. This is apparent
to the eye. The greater width of the jaw in Megaptera and
the greater length in B. musculus are apt to mislead the eye in
regard to curvature.

In regard to *form*, the body has a thicker appearance in
Megaptera on its posterior half, but this is deceptive, being due
to the less development of the upper border towards the coronoid process. The thickness of the body is not greater in
Megaptera than in B. musculus till near the symphysis. The
upper border in Megaptera is gently sigmoid, the convexity on
about the second quarter or more. The distal concavity is
mainly owing to rising up towards the symphysis, on the last
18 inches.

[In *B. musculus* there is very little of the sigmoid character of the
upper border. This is mainly owing to the gradual rising of the
border to the high coronoid process, but there is a slight convexity at
about the second quarter of the border, and a slight rise towards the
symphysis, on about the last 20 inches. The most marked distinction of the body is its greater height on the first quarter in B.
musculus, rising gradually to the high coronoid process. This gives
the body its general tapering form, modified by the slightly sigmoid
upper border.]

There is a difference on the lower border, taking the entire
mandible. Owing to the greater descent of the tuberosity, the
concavity at the lower border of the neck is much more marked
in Megaptera (2 inches deep) than in B. musculus (1 to 1¼ inch
deep).

They differ at the symphysis, on the last 6 or 8 inches. In
Megaptera the lower border is more sloping; at 12 inches from
the end the height is 7 inches in both; at 6 inches, it is the
same in B. musculus, half an inch less in Megaptera.

Foramina.—Both series of foramina present differences. The
internal series in Megaptera, about fifteen in number, open
entirely in the groove. They begin 8 inches in front of the coronoid process, 1½ inch below the border, by two or three close
together, which rapidly form the groove. After an interval of
18 to 24 inches, a series of foramina open at intervals in the
groove, not requiring to ascend to it, except only that the more

posterior open in the lower part of the groove. The groove is not large, wide enough only to lodge the side of the little finger, about $\frac{1}{8}$ to $\frac{1}{6}$ inch deep, and is formed rather of a succession of short grooves continued from the foramina.

[In *B. musculus* the internal series do not begin till about 15 inches in front of the coronoid process, the groove 4 inches earlier. Thereafter there is a continuous groove, formed as in Megaptera, receiving the successive foramina, altogether about 20 in number. There is no wide interval, as there is in Megaptera, between the first group of foramina and those anterior to them. The groove and its foramina are at first rather nearer the border ($1\frac{1}{4}$ inch from it) than in Megaptera, but in B. musculus additional foramina occur farther down. These (2 on the left side, 3 on the right) open at 15 to 23 inches in front of the coronoid process, $1\frac{1}{2}$ inch below the general groove. From each of these a deep groove, wide enough to receive a goose-quill, passes obliquely forwards to enter the general groove, after a course of 7 to 12 inches. The most posterior on the right side again enters the bone, for 3 inches, before it ends in the general groove. These long oblique grooves form a striking distinction of B. musculus from Megaptera.]

The *external series of foramina* are of much greater size, enough to admit a finger somewhat flattened ; the internal series vary in size between that of a goose-quill and crow-quill. In both Megaptera and B. musculus the external series are about 6 in number (6 on the right side, 7 on the left, in both), but in Megaptera they do not begin so soon, and are continued farther forwards. The distance of the first from the coronoid process is, in Megaptera 32 inches, in B. musculus 24 inches. Thus the internal series begin earlier in Megaptera than in B. musculus ; the external series the reverse. The distance of the last of the external series from the symphysis is, in Megaptera 28 to 30 inches, in B. musculus 35 to 38 inches. Owing to their late beginning in Megaptera, the foramina are closer together than in B. musculus. In Megaptera they extend over a distance of 35 to 39 inches ; in B. musculus a distance of 54 to 58 inches. They occur at intervals of 6 to 7 inches in Megaptera, of 9 to 10 inches in B. musculus, but with some variation. They are broader and flatter in Megaptera than B. musculus.

At 13 inches from the symphysis (8 to 10 in B. musculus) the dental canal is unroofed by a fissure $\frac{1}{2}$ inch wide all along in Megaptera, much narrower behind in B. musculus. In both the groove of the internal series of foramina crosses the top of the bone to join the fissure at its posterior end. Thereafter the upper border of the bone is to the inner side of the fissure. The *mental foramen* is incomplete above, and in Megaptera also in front, but is well defined on its floor at about 2 inches from the symphysis. It is larger in Megaptera (height 1¾ inch) than in B. musculus (height 1¼ to 1½ inch). In Megaptera it occupies more than the upper third of the symphysis, in B. musculus scarcely more than the upper fourth.

26. RELATION OF THE MANDIBLE TO OTHER PARTS OF THE SKULL.—When the condyle is placed in contact with the middle of the glenoid cavity, the tip of the coronoid process of the mandible is, in Megaptera, about 15 inches in front of the anterior border of the temporal passage, and about 2 inches in front of the ant-orbital process of the maxillary. Even with the mandible placed thus far back, the coronoid process would quite clear the orbit and ant-orbital process of the maxillary in the closing of the mouth.

[In *B. musculus*, when the condyle is similarly placed in contact with the glenoid cavity, the coronoid process stands below the orbit, behind the middle of it, and could not but strike the malar bone in the closing of the mouth. In order to enable the coronoid process to clear the ant-orbital prominence, the condyle would have to be advanced from the glenoid cavity for about 15 inches. In the 64-feet-long B. musculus I found adhering to the condyle a temporo-mandibular cushion, 30 inches in length, 24 in breadth, 15 to 18 in height (this *Journal*, vol. vi., 1871, p. 123). How much of this lay between the glenoid cavity and the condyle I was unable to note, the mandible having been already detached.]

As the position of the mandible with the condyle in contact with the middle of the glenoid cavity would give a projection at the symphysis of only 2¼ inches beyond the beak (and would in B. musculus place it 2 inches short of the beak), it is evident that there is naturally a considerable interval between the glenoid cavity and the condyle. If the temporal muscle goes to the coronoid process in Megaptera it will, after turning round the anterior boundary of the temporal passage, have a long way to travel, and that by passing close below the soft parts of the orbit. Dissection of the muscles and other soft parts here is necessary to the interpretation of the bone at and near the coronoid process.

27. HYOID BONE.—TABLE III. *Measurements of the Hyoid Bone,* in inches :—

	Megaptera.	B. musculus, 50 feet long.
1. Extreme width,	$22\frac{1}{4}$	27
2. At middle of body, breadth (ant.-post.),	5	$5\frac{1}{2}$
3. ,, thickness,	$1\frac{3}{8}$	$1\frac{3}{4}$
4. Breadth of great horn, at middle,	3	$3\frac{1}{4}$
,, on outer end,	$2\frac{1}{2}$	$2\frac{1}{2}$
5. Thickness of ditto, at middle,	$2\frac{1}{4}$	3
6. ,, on outer end,	2	$1\frac{7}{8}$
7. Stylo-hyal, length,	12	14
8. ,, greatest breadth,	$2\frac{1}{2}$	$2\frac{1}{4}$
9. ,, thickness at same part,	$1\frac{1}{2}$	$1\frac{1}{2}$
10. ,, both ends,	$1\frac{3}{4} \times 1$	...
11. ,, thick end in B. musculus,	...	$2\frac{1}{2} \times 1\frac{3}{4}$
12. ,, flat end in ditto,	...	$2\frac{1}{2} \times \frac{7}{8}$
13. Weight of united body and great horns, in ounces,	$48\frac{1}{2}$	73
14. ,, of the stylo-hyals together,	$20\frac{1}{2}$	$30\frac{1}{2}$

28. CHARACTERS OF THE HYOID BONE, IN COMPARISON WITH THOSE OF B. MUSCULUS.—Although the hyoid bone of B. musculus strikes the eye as larger and more massive than that of Megaptera, it would not be so were we to go merely by the comparative length of the two skeletons. But there are differences in form.[1]

Both ends of the stylo-hyals are flat in Megaptera, while in B. musculus the ends differ materially in thickness, as seen in the measurements given in the Table. The anterior conical processes, commonly regarded as outgrowths from the body, but

[1] The hyoid of Megaptera is at once known from that of *B. borealis* by the broad stylo-hyals of the latter ; also by characters of the body and great horn, especially by the flatness of the great horn in B. borealis. In this 35-feet-long B. borealis the stylo-hyals have not attained the breadth figured by Flower, from the Leyden (Java) finner (*Proc. Zool. Soc.*, 1864, p. 406), and figured, from the same skeleton, by Van Beneden and Gervais (*loc. cit.*, pls. xiv., xv. fig. 28, B. Schlegelii). The vertebræ and other characters figured in that plate leave no doubt that their B. Schlegelii is the same species as my B. borealis. The stylo-hyals in my B. borealis are liker those figured by Van Beneden and Gervais (pls. x., xi. figs. 15, 16, "d'après le squelette de Leide") in their so-called B. laticeps, except that in mine the lower end is more pointed. In this 35-feet-long B. borealis the greatest breadth of the stylo-hyals is 4 inches ; length 13, the broad part $9\frac{1}{2}$, the tapering lower part 5 inches.

As Dr Gray's New Zealand species is founded on only one specimen, and that of the ear-bone only, it is evident that further observations are required on the form of this part of the skull of the **Megaptera** of the Greenland seas and of other seas. It has to be determined whether the supposed specific difference, remarkable as it is, may not be a matter of age, or of non-union of the two parts from which this portion of the ear-bone is developed, or of variation within the same species, or possibly of the mutilation of the thinner part in detaching the bone. But should this wing-like expansion of the periotic bone prove to be diagnostic of the Megaptera of the South Pacific, then it must be held that we have here a specimen of the Megaptera longimana of New Zealand migrated thus far northwards.

PRINTED BY NEILL AND COMPANY, EDINBURGH.

For Explanation of Plates I. and II., see page 17.

For Explanation of Plates III., IV., and V., see pages 59–60.

For Explanation of Plate VI., see pages 144–145.

Note in regard to fig. 16, Plate V. :—The <u>limits</u> of the Anterior and Posterior Cartilages of the Pelvic Bone are <u>not marked</u> on the figure. The length of these Cartilages is given in Table V., Part II. page 50.

The limits are marked off by the dotted red lines, since added.

fig! MEGAPTERA LONGIMANA.

(FROM A PHOTOGRAPH, AT STONEHAVEN, near ABERDEEN, 1864.)

LENGTH 40 FEET.

Fig. 2. ¹⁄₃₆

Fig. 3, ¹⁄₃₆

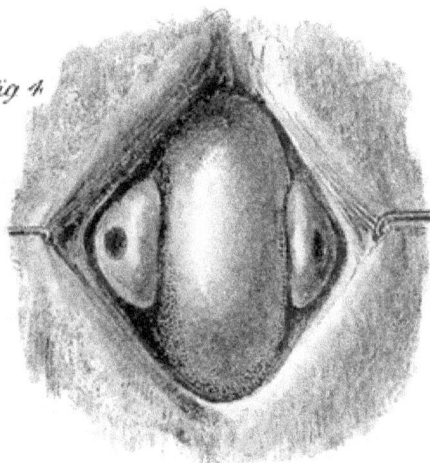

Fig. 4.

Mammillary Pouch and Nipples
of Male Megaptera.
(NATURAL SIZE)

Fig. 5, ¹⁄₄₀

John Struthers. Delt.

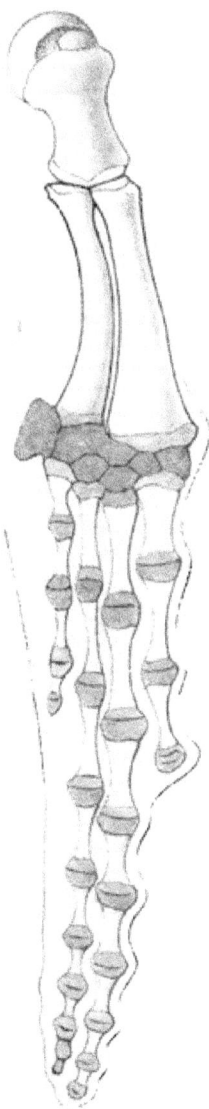

Fig. 7.
1/24

Left Scapula.
OUTER SURFACE.

Fig. 8.
1/24

Glenoid Cavity
& Coracoid.

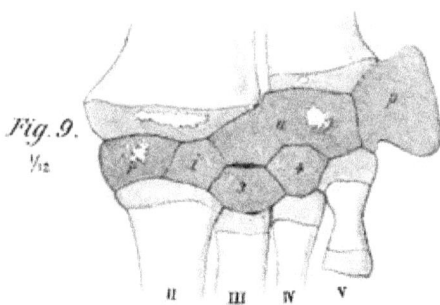

Fig. 6.
1/24.

Pectoral Fin.
(LEFT) INNER SURFACE.

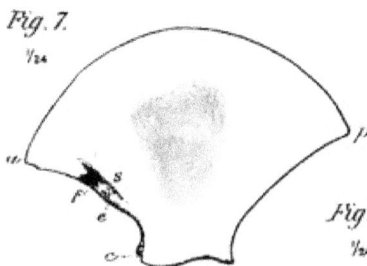

Fig. 9.
1/12

Left Carpus, &ᶜ
(SECTION) DORSAL VIEW.

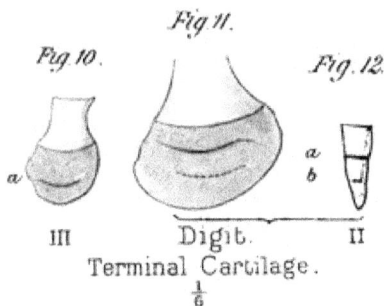

Fig. 10.

III

Fig. 11.

Digit.

Fig. 12.

II

Terminal Cartilage.
1/6

MEGAPTERA   LONGIMANA.

John Struthers Delᵗ

Andrew Gibb & Co., Lithographers.
3. Queen-Street, Aberdeen.

Journ. of Anat. & Phys. Jany 1888.

Vol. XXII.
N.S. Vol. II. } *Pl. IV*

FIG. 13.

FIG. 14.

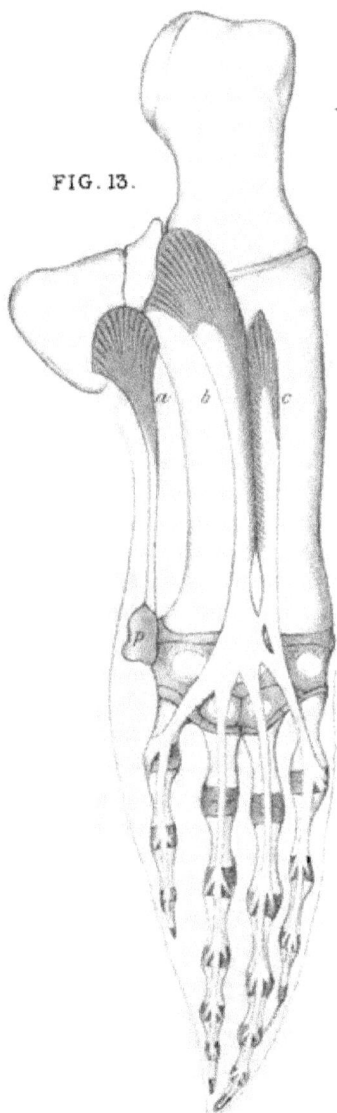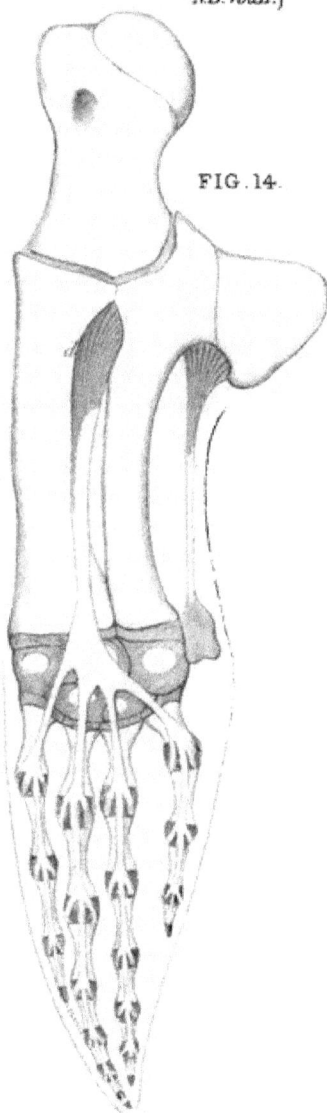

Flexor and Extensor Muscles of the Fingers, ½.

FIG.13. Inner aspect *a* Flexor carpi ulnaris   *p.* Pisiform cartilage
        b F. digitorum ulnaris. (F. profundus digitorum)
        c. F. digitorum radialis (F. longus pollicis.)
FIG.14. Outer aspect *d.* Extensor communis digitorum.

BALÆNOPTERA MUSCULUS.

64 Feet Long.

John Struthers, delt

Andrew Gibb & Co. Lithographers.
3. Queen Street. Aberdeen.

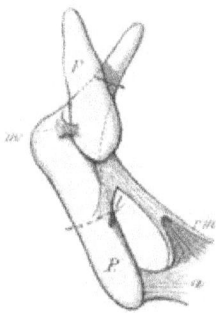

Journ. of Anat. & Phys. Jany 1888.　　　　　Vol. XXII.⎱
　　　　　　　　　　　　　　　　　　　　　　N.S. VolII.⎰ Pl. V

FIG 16.
PELVIC BONE
Femur, &c.
of
Megaptera
longimana.

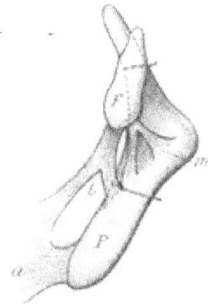

(FIG. 16. Right.)　　　　　　　　(FIG 16 Left)

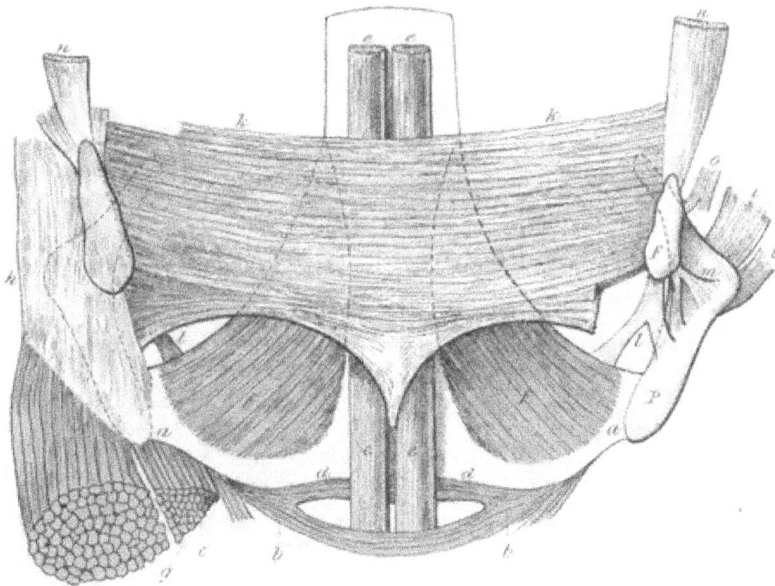

FIG 15. Ligaments Muscles, &c connected with
Pelvic Bone and Femur of
MEGAPTERA　LONGIMANA.

John Struthers. delt.　　　　　　　　Andrew Gibb & Co., Lithographers
　　　　　　　　　　　　　　　　　　　3. Queen Street, Aberdeen.

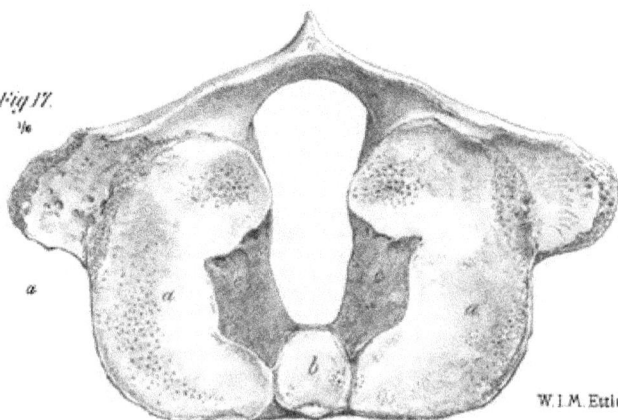

FIG. 17. Atlas of Megaptera Longimana, hinder aspect
  a a Lateral articular surfaces.
  b Mesial articular surface.
  c c Ligamentous Area.

W.I.M.Ettles, Del.

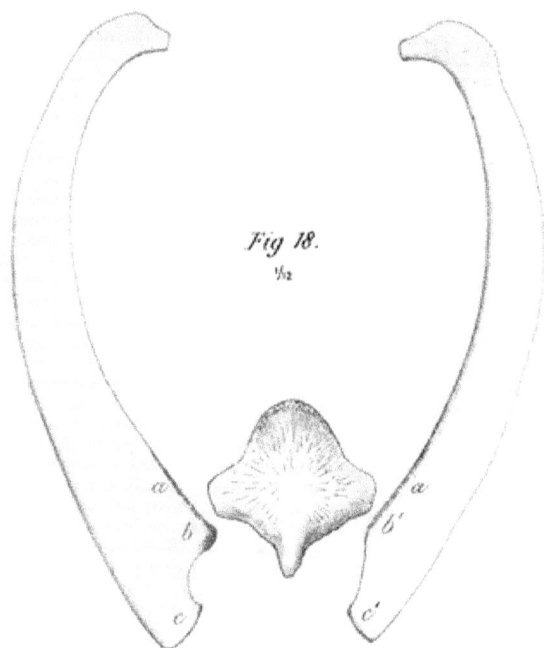

FIG 18. Sternum and First Pair of Ribs, anterior aspect
  a a. Rough Part of costal border.
  b Anterior, and *c* Posterior angle of end of Right Rib
  b' and *c'* Same parts of Left Rib showing the differences